The Uses and Methods of
GAMING

The Uses and Methods of
GAMING

Martin Shubik

ELSEVIER

New York / Oxford / Amsterdam

ELSEVIER SCIENTIFIC PUBLISHING COMPANY, INC.
52 Vanderbilt Avenue, New York, N.Y. 10017

ELSEVIER SCIENTIFIC PUBLISHING COMPANY
335 Jan Van Galenstraat, P. O. Box 211
Amsterdam, The Netherlands

Library of Congress Cataloging in Publication Data

Shubik, Martin.
   The uses and methods of gaming.

   Includes bibliographies and index.
   1. Game theory.   I.   Title.
QA269.S56   519.3   75-8275
ISBN 0-444-99007-0

Manufactured in the United States of America

To

MY WIFE, JULIE,

MY DAUGHTER, CLAIRE LOUISE, *and*

MY SISTER, IRENE,

*none of whom have the slightest intention
of really reading this book.*

Contents

Acknowledgments   ix

Introduction   3

1   Gaming, Simulation, and Theories of Behavior   6

2   On the Many Goals of Gaming   26

3   Techniques, Modeling, and Languages   49

4   Costs and Procedures   59

5   Facilities   83

6   Intention, Specification, Control, and Validation   100

7   A Guide to Information Sources on Gaming and
    Related Topics   119

8   Gaming for Business, Management, Operations
    Research, and Economics: A Literature Guide   134

9   Experimental Gaming: A Literature Guide   144

10 Gaming in Political Science, International Relations
    and for Military Purposes: A Literature Guide   166

11 Gaming and Related Topics: A Literature Guide   182

    Index   193

Acknowledgments

The author wishes to acknowledge the help and assistance given to him in many interesting discussions by his colleagues at Rand and elsewhere. In particular, I am indebted to Gary Brewer, Norman Dalkey, Harvey DeWeerd, Herbert Goldhamer, Olaf Helmer, William Jones, Edward Paxon, and Milton Weiner. My acknowledgments are for help, not merely with this book but with its companion book, *Games for Society, Business and War*.

I am also indebted to several other old friends and colleagues. In particular, to Austin Hoggatt, with whom I have had many conversations on the uses and potentialities of experimental gaming; to Gerald Shure on the design of laboratories for experimentation; to Robert Noel and Harold Getzkow on the uses of gaming in international relations, and to Anatol Rapoport for many stimulating discussions on experimental games and the stage of the universe.

I owe a particular debt of gratitude to my secretary, Mrs. Elizabeth Walker whose unfortunate task involved endless footnoting, checking, rechecking and retyping what appeared to be a never-ending amount of manuscript.

I would like to thank Daniel Couger for permission to quote from a survey of his, Robert Noel for permission to quote from an unpublished description of his laboratory, and Austin Hoggatt, J. Esherick, and J. T. Wheeler and the *Administrative Science Quarterly* for permission to reproduce a diagram of the Management Science Laboratory.

The Uses and Methods of
GAMING

Introduction

This book is designed to present a brief sketch of the many types and purposes of gaming together with a instructions for those who wish to construct games for teaching, operational, experimental, and other purposes.

Because many different disciplines are involved, it is desirable to have a background in the basic theory underlying work in operations research, social psychology, the theory of games, simulation, and several other disciplines. No attempt is made to provide this background here. In a related publication, *Games for Society, Business, and War*,[1] the necessary game theoretic background has been provided. In Chapters 7 to 11 of the present work, the interested gamester will find a guide to all the reading that he needs to obtain both a theoretical background and a methodological approach to gaming.

The activity of gaming may broadly be described as the using of a game for purposes of teaching, training, operations research, therapy, or entertainment. The game frequently but not always provides a simulated background for the player. In Chapter 1 the meaning of gaming is discussed in greater depth, and the relationship among gaming, simulation, and various theories of behavior is explored. A brief sketch of some of the conceptual difficulties which may trap the unwary is given. The literature guide for a more detailed discussion is given in Chapter 7.

The next five chapters deal with gaming methodology. In Chapter 2 a broad sketch of the different purposes of gaming is given. Comments on the uses of gaming made in Chapter 1 are expanded. Chapter 3 provides a brief sketch of techniques and problems in simulation and modeling for the construction of gaming exercises.

Chapter 4 and 5 deal with gaming costs and facilities. Gaming can easily be extremely expensive. It is difficult to determine the exact expenditure on games utilizing joint facilities and free resources, which are usually not counted. Although one might argue that free resources should not be counted, a comparison of the cost of games requires data about all of the resources used. A resource that may be free in one instance may not be free in another. For example, for games run by the military and games run at universities, frequently the facilities are supplied free of charge. However, if a third party wished to replicate a game he might have to hire the facilities.

Some games involve a considerable expenditure of time by top personnel. For example, four-star generals or high-ranking diplomats may be used in certain operational games. In games played at universities, members of the faculty may give many hours of their time. This will not be recorded as a cost.

Whether one is using gaming for operational, teaching, or experimental purposes, the need to replicate is imperative. No matter what criterion for validation one may wish to apply, one's level of confidence will depend considerably on having obtained comparable results under comparable conditions. It is desirable to be in a position both to replicate a game and to compare the outcomes. Few games have been carefully replicated and among those where replication has taken place, few have had the obtained data subjected to careful comparative analysis.

The ability to record games by such means as audio tapes, video tapes, and computer tapes allows virtually all of the trivia from any game to be retained. The net result is that it becomes easy to have inventories of undigested data, which in the long run will probably be thrown out when the individuals who gathered them finally realize that they do not know what to do with the information they possess.

With the growth of gaming and the need for both operational and experimental game facilities, there have been relatively large-scale efforts devoted to the construction of gaming laboratories. In Chapter 4 some observations are made on the major problems and difficulties to be encountered in the construction of such facilities. Included within this chapter are also some general comments on the methodology of gaming showing how standardization can affect the costs and value of a game.

Chapter 6 deals with validation. Each special, different use calls for a different concept of validation. The goals of the sponsor, the designer, the controllers, and the players may all be different. Thus, we have the additional problem of deciding upon "validation for whom?" since the concept means something slightly different to each user.

It is extremely important to sort out stated from unstated purposes for gaming and stated from unstated criteria of validation. The successful use of a gaming exercise depends heavily upon the prior clear understanding of purpose combined with a preliminary decision on what measures of success are to be applied.

The remaining chapters contain a guide and partially annotated bibliography to literature on the many different applications of gaming. Only a few specific games are discussed here. There are now literally thousands of articles and books on various applications of gaming. They range from pure salesmanship to competent scholarship, and the criteria for acceptability vary heavily from field of application to field of application.

REFERENCES

1. Shubik, M., *Games for Society, Business and War*, Amsterdam: Elsevier, 1975.

1 Gaming Simulation and Theories of Behavior

The major use of gaming the world over is by the military for teaching, training, and operational studies. Probably the second most visible formal use of gaming is in teaching, training, and experimentation with business and economic games. In the past few years there has also been a considerable growth in the use of games for studying social and other problems. These include land-use games, and games that otherwise involve political and social processes. At another level, there are many activities in elementary schools that could be classified as games.

The major expenditures in gaming activities (such as sports) clearly are for entertainment. In general, we do not consider these purely recreational games when discussing teaching, operations, or experimentation. It is the author's contention that sports, parlor games, and other allied activities, such as gambling, present the behavioral scientist with a potentially rich source of observations. That it has not been tapped is due to difficulties in data gathering and to a prejudice by many scholars that such an occupation would be frivolous.

Gaming entails the use of a scenario, game, simulation, or model to provide a background or environment in which a set of individuals usually referred to as the players will act. The environment is almost invariably a simulation or model of a real environment. Thus, for example, in a business game, frequently there is a computer model that supplies the representation of the firms and the industry; in certain military exercises a sand table may provide a model of the terrain. In some instances, however, the game takes place in an environment that is not a model of something else. The chess board could be regarded as a model of a battlefield; yet it is sufficiently abstract and formal that chess can best be described as a model of itself. The same clearly goes for most card games, such as bridge or poker. Although one may draw analogies between poker and life, the formal rules of the game do not pretend to model life.

In a game, the players either play simulated roles or they play themselves. For example, one might have a business game in which a student is instructed to behave as though he were the president of General Motors. In this instance, he is playing a simulated role. In a military exercise a major may either be required to simulate the role of a general, or to play his own role of a major. Chess players simulate no one; they merely play chess to their own account.

If one wished to study the effects on morale of the nature of fighting in an atomic war, the scientifically most accurate way would be to observe an atomic war in progress. For obvious reasons, the study of a simulation of the phenomenon, even though it undoubtedly will be less accurate, is more desirable than a study of the actual event, and, if done with care, should provide some insight about what might happen in the real situation.

SIMULATION, GAMING, AND GAME THEORY

It is desirable to understand the differences involved in gaming, game theory, and simulation. Even though there is a tendency to use these words interchangeably, there are considerable differences in usage among the various practitioners of gaming.

Simulation

A simulation involves the representation of a system or an organism by another system or organism that purports to have a relevant behavioral similarity to the original system. The simulator is usually simpler than the system being simulated and is more amenable to analysis and manipulation.

A simulation is a model. The individual interested in employing a simulation usually has a specialized set of questions about the behavior of the system. As such, he wants the simulation to emphasize the salient features he has studied. Misplaced realism may easily lead him to build a more complex and unmanageable representation of the system he wishes to study. Simplicity, relevance, and the appropriateness of the aggregation and abstraction are the key elements to successful model-building for simulations.

7

In general, both in the behavioral sciences and in management applications, simulations being built currently tend to utilize a high-speed digital computer. Board games, sand tables, and analog computer models provide other examples of nondigital computer simulations.

There are two crude categories of simulation of interest to the gamer. They can best be called "tactical simulations" and "strategic or exploratory simulations." The distinction is both a quantitative and a qualitative one. Under the category of tactical simulations comes much of the work done in operations research, including the simulation of production-inventory scheduling, waiting-line problems, and traffic-flow models. The system to be simulated is relatively well-defined, and its components can be accurately described and mathematically modeled in a satisfactory manner. The modeler can be reasonably content that he is able to produce a relevant, parsimonious model of the system he wishes to study.

Strategic or exploratory simulation involves large-scale models of economic, social, or political phenomena where the size and complexity of the models coupled with difficulties in measurement call for judgment or estimation. Examples range from the large-scale econometric models, such as that of the Brookings Institute, where at least most of the variables are reasonably well-defined and where econometric methods have been used to try to estimate parameters, to the models suggested by Forrester[1] or to the scenarios utilized in political-military exercises.

Such problems as pollution, social welfare, quality of life, and national morale do not lend themselves to easy classification. Large-scale models wh:ch have them as variables require that the user be extremely cautious before drawing analogies or claiming validation. The setting for a play is a simulation, as is the scenario for a political-military exercise. Such a scenario is soft, environment-rich, and vague, in contrast with the "hard," environment-poor and precise model offered by a tactical simulation of a transportation system or logistics system.

Gaming

Gaming, in contrast to simulation, of necessity employs human beings in some role, actual or simulated. A gaming exercise may employ human beings acting as themselves or playing simulated roles in an environment which is either actual or simulated. The players may be experimental subjects being observed or they may be participating in the exercises for teaching, training, or operational purposes.

When playing chess, neither the role of the player nor the environment is simulated. When students are told to assume the roles of top decision makers in a computerized business game or war game, the roles of both the players and the environment are simulated.

Game Theory

Occasionally gaming and game theory are confused. The first, as has already been noted, is a method employed in experimentation, operations or teaching. The second is part of a growing body of theory concerning decision making. In particular, game theory provides the language for the description of conscious, goal-oriented decision-making processes involving more than one individual. It has provided the tools for analysis of certain relatively subtle concepts, such as the state of information, the meaning of choice, move, strategy, and payoff.

Game theory is a mathematical theory which can be and has been studied as such, with no need to relate it to behavioral problems or to games *per se*. It can, however, be given either a normative or descriptive interpretation in application to multiperson decision-making. In general, game theoretic reasoning and analysis are of considerable use in constructing, discussing, and analyzing gaming exercises. A brief sketch of the features of game theory relevant to the gamer is given below. However, a more exhaustive and relatively self-contained exposition is given elsewhere.[2]

RIGID AND FREE-FORM GAMES

Games can be classified not only according to their purpose but according to the formality of their rules. In particular, some games are "free-form" while other games have extremely rigid rules. In a free-form game many of the rules are made up as the game proceeds. The simulated environment is purposely left somewhat unstructured. Not everything is completely defined in advance. It is evident that such a game is not an immediate candidate for complete computerization. For example, it would be extremely unwise to try to computerize an international relations game without first playing it considerably in a loose, noncomputerized form until the model builders are satisfied that they know how to isolate the phenomena they wish to model.

In contrast, a rigid game has all of the rules completely specified and well-defined in advance. Usually such a game can be computerized, and its play depends far more heavily upon numerical than verbal or other forms of information. Many operations research games, technical war games, and business games tend to be rigid-rule games.

Free-form games tend to be "environment rich" or loosely defined and imaginative, whereas rigid-rule games are frequently "environment poor" with a minimum of vagueness, little verbal description of basic importance and with little room for radical reinterpretation.

A game may be regarded as environment poor if there is little if any question that all players are fully able to comprehend all of the rules. In such a game the number of variables must be highly limited and the number of, and details concerning the rules must be few. In contrast, in an environment-rich game considerable attention may be paid to institutional detail, to context, and to obtaining a realistic and rich scenario.

There is a continuum of cases between environment-poor and environment-rich games. Nevertheless, the dichotomy calls our attention to two extremely different approaches. For the most part, the type of gaming exercises used in experimental psychology tends to be environment poor, while many experiments in social psychology and most political science gaming tend to be environment rich.

Environment-rich games frequently involve much interpretation of the rules of the game as well as its play. Ingenuity and imagination may be required in playing a rigid-rule game, such as chess, but the imagination is required for the play, not for the interpretation of its rules.

THE USES OF GAME THEORY
IN MODELING OF GAMES

A way in which we may understand both the power and the limitations of the uses of game theory is to consider its potential role in the modeling of gaming exercises. There are six fundamental problem areas which are of concern to the model builder or game constructor and which may be the source of considerable difficulties when he attempts to produce an adequate model for the purpose he has at hand. They are:

1. the definition of rules and problems of wording and coding
2. the definition of rules, the meaning of rationality, and problems of information and data processing ability
3. the specification of payoffs, goals, and motivation
4. the specification of rules for environment-poor and environment-rich games
5. the meaning of rationality and the concepts of solution for multiperson games
6. the specification of players as individuals or groups

Before discussing these problem areas, a brief discussion of game theory is presented in order to provide a frame of reference to compare abstract games with real games, and games for playing with actual social, economic, or political processes. A more complete background is found in Chapters 2 to 6 of *Games for Society, Business and War*.[1] Alternative references are also provided in Chapter 7 of the present work. In this section, a checklist of the basic features of a game as viewed through the methodology of game theory is given. In the subsequent section, what game theory leaves out and what other behavioral theories encompass are noted.

Rules and Descriptions of Games

The theory of games offers three major formal descriptions of a game, all of which are of importance and use to the game designer or user for different purposes. They are:

1. the extensive form
2. the strategic or normal form
3. the coalitional form

The first of these descriptions lays stress upon details of information and the fine structure of the process of a game, the second concentrates on strategic considerations, and the third stresses coalitions and the combinatorics of alliances, bargaining, and group formation.

The Extensive Form

Consider a simple game where Player 1 moves first and has to make a selection between two alternatives. After he has made his choice, Player 2 is called upon to move but is not informed about

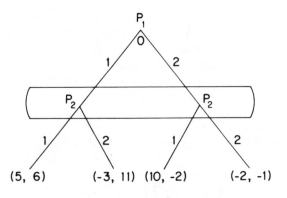

FIGURE 1.1

Player 1's choice. He has to select between two alternatives. After both players have chosen their moves the game is over, and, depending upon what they have done, they receive their payoffs. Figure 1.1 provides a diagram of this game. Each node of this diagram (known as a game tree) is labeled with the name of its "owner." Thus, the first node has the label P_1 indicating that the first player must make his choice (the first node also has the label 0 to indicate that it is the first node). The second level of nodes associated with branches 1 and 2 both bear the label P_2 indicating that it is the choice of the second player to make his move. Both of these second-level nodes are enclosed in a curve known as an information set. This indicates that when Player 2 is called upon to move, he is unable to distinguish among the nodes included in a single information set. In other words, this is a way of representing his ignorance of what the other player has done.

Given that each has two alternatives, there are four possible ways in which this game could be played. The payoffs resulting from any one of these plays are indicated by the pair of numbers at the end of each terminal branch. In each case the first number is the payoff to the first player and the second number the payoff to the second player.

In theory, at least, games such as chess, poker, or a business game could be laid out in this manner. Considerable ingenuity must be exercised in describing the complex information conditions that may exist in some games. However, the game tree can map out the conditions if the meaning of what constitutes a move is clear and our understanding of the information conditions is unambiguous.

There has been a certain amount of simple experimental work using the game tree representation of a game. Because game trees can fast become unmanageably large and unwieldy, they are rarely used in examining large games. However, in general, the main use of this representation to the gamer is as a device for checking his understanding of information conditions and the description of moves in a game he may wish to design.

The Strategic or Normal Form

The representation of a game used most frequently in gaming experiments, especially in social psychology, is the matrix game. The same game that has been described in Figure 1.1 could be somewhat differently described in Figure 1.2.

This extremely simple game is illustrated by a 2 × 2 matrix. The numbers in the cell present the payoff respectively to Players 1 and 2 as a result of the combinations of their choice. Although most of the experimentation has been with 2 × 2 matrix games, the formal description of a game holds for any number of players, each with any number of strategies. For example in Figure 1.3, a three-person matrix game where each player has to make a choice between two alternatives can be represented by two matrices where each cell contains three entries. The entries are the payoffs to the first, second, and third players, respectively. As is shown in Fig-

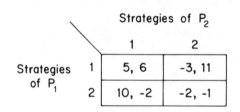

FIGURE 1.2

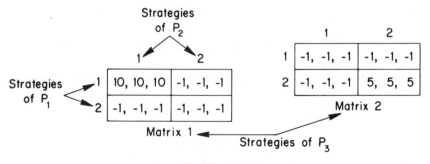

FIGURE 1.3

13

ure 1.3, Player 1 chooses the row, Player 2 chooses the column, and Player 3 chooses between the two matrices.

When we consider Figures 1.1 and 1.2 the difference between a move and a strategy cannot be seen. We must consider a slightly more complicated case in order to make this distinction. Suppose that after Player 1 had made his choice Player 2 were informed of this choice before he had to act. In this instance, Player 2 has more information. The two nodes that in Figure 1.1 were enclosed in a single information set now will be closed in separate information sets as is shown in Figure 1.4.

The effect of this additional information on the number of strategies available to Player 2 can be seen as follows. Suppose that Player 2 had to delegate his authority to an agent to play for him. We might regard as a strategy a complete book of instructions to be given to the agent to cover all contingencies. If the agent has no information about Player 1's choice prior to making his choice, there are only two sets of instructions which can be met. They are either to, select Move 1 or to select Move 2, as was portrayed in Figure 1.1. If, on the other hand, his agent has information concerning Player 1 before he makes his choice, there are four books of instructions which can be met. They are as follows:

1. If Player 1 selects his first move, choose 1; if he selects his second move, choose 1.
2. If Player 1 selects his first move, choose 1; if he selects his second move, choose 2.
3. If Player 1 selects his first move, choose 2; if he selects his second move, choose 1.

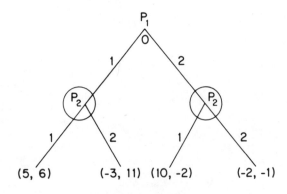

FIGURE 1.4

Strategies of P_2

Strategies of P_1		(1, 1; 2, 1)	(1, 1; 2, 2)	(1, 2; 2,1)	(1, 2; 2, 2)
	1	5, 6	5, 6	-3, 11	-3, 11
	2	10, -2	-2, -1	10, -2	-2, -1

FIGURE 1.5

4. If Player 1 selects his first move, choose 2; if he selects his second move, choose 2.

The strategic or normal form of this game is a 2 × 4 matrix as is shown in Figure 1.5.

A strategy may be regarded as a plan which covers all contingencies given the information available to the player. It should be rather obvious that even in a simple game, such as tick-tack-toe, there are millions upon millions of strategies. However, the way humans actually play the game is that they aggregate them or select among them according to various heuristics ("rules of thumb") or behavioral rules.

The Coalitional Form of a Game

Suppose that we were not particularly interested in the details of what strategies the players actually select but were more concerned with having the players come together, discuss matters, argue, bargain, collude, or cooperate about how they intend to actually play the game, and to decide upon how to split the proceeds. In the process of doing so, of course, all individuals and subgroups may wish to use as arguments in their bargaining, their threats, and the payoffs that they can enforce if they fail to cooperate as a single unit.

The gains to every coalition can be represented by means of a set function called the characteristic function of a game. It is a set function in the sense that it is defined over all possible sets of players. Thus, for a game with 10 players it will have 2^{10} or 1024 values. In general, for a group of n players there are 2^n coalitions which can form. (This includes *pro forma* the coalition consisting of no one. Hence, if we wish to exclude this, we must subtract 1 from the number of coalitions.) The coalitional or characteristic function form for the game shown in Figures 1.1 and 1.2 is shown

15

in Table 1.1. The coalitional form of the three-person game shown in Figure 1.3 is shown in Table 1.2.

TABLE 1.1

$$v(\theta) = 0$$

$$v(1) = -2, \quad v(2) = -1$$

$$v(1, 2) = 11$$

TABLE 1.2

$$v(\theta) = 0$$

$$v(1) = v(2) = v(3) = -1$$

$$v(1, 2) = v(1, 3) = v(2, 3) = -2$$

$$v(1, 2, 3) = 30$$

In both Tables 1.1 and 1.2 we have formally assigned a worth of zero to a coalition of no one. For purposes of exposition here, we could have left this out, but in general it is safer to include it. In Table 1.1, Player 1 acting by himself can guarantee no more for himself than -2. Similarly, Player 2 can guarantee no more than -1. If they act together they can obtain a total of 11. In the three-person game, no single player can guarantee more for himself than -1. No pair of players can guarantee more for themselves than -2. However, if all three are willing to act together, they can obtain as much as 30.

Implicit in this description has been the assumption that the units of payoff have the same value to all players and that they can work out sidepayments and all types of settlements among themselves. In some instances, this is not the case. For example, it is explicitly forbidden by law for colluding firms to make payments to each other; this, of course, does not stop collusion. It merely makes the process of collusion much more complicated as the sidepayment mechanism and the *quid pro quo* can no longer be expressed by a single number.

In particular, if sidepayments are not possible, the structure of threats among the players may become quite complex. For example, if a wealthy parent is forbidden by law to pay ransom money for his child, the threat possibilities for a kidnapper are considerably changed from the situation in which the payment of blackmail or ransom money is legal.

16

Outcomes, Payoffs, and Sidepayments

In game theoretic modeling two important levels of distinction are made, each of which is of considerable applied value to a gamer engaged in virtually any type of gaming. They are the distinction between outcomes and payoffs in a game, and the distinction between payoffs and the various types of sidepayment possibilities that exist in different games.

In the payoff matrices drawn in Figures 1.2 and 1.3, numbers representing the payoffs were given for each outcome. A confusion between payoffs and outcomes is easy to make. An outcome is the result of the play of a game, not what that result is worth. For example, in chess, the outcome is white checkmates black; black checkmates white; or the game is a draw. It is usually assumed that white's preferences over the outcomes are such that he prefers a win to a draw and a draw to a loss. However, if white is playing with his boss and has just beaten him four or five times, it well may be that he prefers to lose. In both experimental and operational gaming it is extremely important to make sure that the relationship between outcome and payoffs for all players is absolutely clear and explicit.

Even if the payoffs are well-defined to each individual, it is not always the case that individuals necessarily can easily compare payoffs or can make transfers among themselves. When there is only money at stake, it may be possible to do so relatively easily. However, when power, prestige, and other social factors are involved, simple assumptions concerning comparability of goals and ease of sidepayments cannot be made. The game theorist is furthermore aware of the subtleties of form which sidepayment mechanisms take both in theory and in practice. *Quid pro quo* has many manifestations.

On Solutions

There are three broad categories of solution of interest to the game theorist, and they are all relevant to the gamer. They are:

1. noncooperative solutions
2. cooperative solutions
3. dynamic and behavioral solutions

What is meant by a solution to a game? A good model in and of itself may almost be a solution in the sense that it helps us to understand the phenomenon we are examining. However, in general, a solution can be regarded as a prescription or a prediction that a certain outcome or set of outcomes will result from the play of the game by a set of individuals. Some solution concepts may be rather broad and have weak resolution power. For example, a solution may merely predict an optimum outcome without giving any information whatsoever concerning how the payoffs will be distributed. Other solutions may predict or prescribe a single point. For example, solutions concerned with fair division and equity will in general tend to call for a specific way in which the payoffs should be distributed.

There is a considerable literature on both noncooperative and cooperative solutions together with many variations. References are given later for those who wish to study them in further detail. Most of the developments in theory of games have been with static theories of solution. A unified dynamic theory of games has not been achieved. However, it is precisely at the most important level which calls for the understanding of dynamics that both game theorists and the behavioral scientists pool ignorance; for none of them appears to offer more than some limited explanations of behavior over time. Even so, the gamer has much to learn from an understanding of the static solutions developed by game theory. In particular, an important distinction must be made between two-person, zero-sum games and other games which are nonconstant sum or which have more players. (Technically, the distinction should be made for two-person, constant-sum games which are strategically equivalent to zero-sum games, but have a constant added to the payoff of each player for every outcome.)

In a two person, zero-sum game, the amount that one player wins is precisely the amount that the other player loses. In Figure 1.6a an example of a two-person, zero-sum game is given. It is displayed in a manner similar to the display in Figure 1.2, but because we know that the payoff to Player 2 is the negative of the payoff to Player 1, the additional information is redundant and we might as well represent it merely by the payoff to the first player as shown in Figure 1.6b.

The two-person, zero-sum game has as its famous normative solution the maxmin strategy suggested by von Neumann. Player 1 can examine each row and consider what is the largest amount that he can obtain, given any action of his opponent. If he plays the first row, the worst that can happen is a payoff of 2. If he plays the second strategy, the worst that can happen is a payoff of 1. The maximum of the minima occurs if he were to play his first strategy.

18

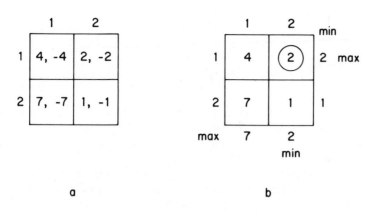

FIGURE 1.6

Similarly, Player 2 wishes to minimize the payoff to Player 1, as that is the equivalent of maximizing the payoffs to himself. Using this reasoning, if he selects the first row, Player 1 could obtain as high as 7. If he selects the second row, Player 1 could obtain as high as 2. If he wishes to minimize the maximum obtainable by Player 1, he will select his second strategy. In the terminology of game theory, this game has a saddlepoint determined by the strategy pair of 1 and 2 for Players 1 and 2. It is easy to see that even if there were a communication channel between the players, it would essentially serve no purpose. As their interests are diametrically opposed, the players really do not have anything to talk about. This is not the case with nonconstant sum games.

In the game illustrated in Figure 1.2 and in coalitional form in Table 1.1, we observe immediately that although there is a conflict of interest about who gets what piece of the pie, there is a large community of interest concerning how big a pie they will share. If they jointly agree to use their first strategy, they have 11 to share between them. If they both use their second strategy they will end up with losses of – 2 and – 1, respectively.

The most considered solution concept suggested for nonconstant sum games where there is limited communication between the players is the Nash noncooperative equilibrium solution. It is described for the two person case but it clearly generalizes for any number of individuals.

Let s_i be a strategy for Player i. In order to denote a specific strategy, as contrasted with a general strategy, we use the notation $\bar{s}_i$. A point $(\bar{s}_1, \bar{s}_2)$, is an equilibrium point if it satisfies the following conditions:

19

$$\text{Max } P_1(s_1, \bar{s}_2) \qquad \text{implies } s_1 = \bar{s}_1$$
$$s_1$$

$$\text{Max } P_2(\bar{s}_1, s_2) \qquad \text{implies } s_2 = \bar{s}_2$$
$$s_2$$

where $P_1(s_1, s_2)$ and $P_2(s_1, s_2)$ are, respectively, the payoff functions for Players 1 and 2. Formulated in this manner the solution is static and gives us no insight into how the dynamics of play might bring it about.

An examination of the nonconstant sum game in Figure 1.2 reveals that it has a unique equilibrium point at the strategy pair (2, 2) where Player 1 loses 2 and Player 2 loses 1. This game is a nonsymmetric example of the Prisoner's Dilemma. A symmetric example can be obtained by subtracting 1 from each of the payoffs to the second player.

In the three-person game illustrated in Figure 1.3 there are two noncooperative equilibria—one when the strategy triad (1, 1, 1) is played and another when the strategy triad (2, 2, 2) is played. The first one yields payoffs of 10 to each player and the second one yields payoffs of 5 to each player. Nevertheless, in the following sense, they are both equilibria. Suppose that you were playing in this game and you had been told that all previous players had selected their second strategy; furthermore, you are not able to communicate with the other players. If you unilaterally were to shift to your first strategy, you might easily bring negative payoffs to everyone, including yourself. Thus, the mechanism for shifting from Strategy 2 to Strategy 1 for everyone is not really there without communication.

There are other noncooperative solution concepts with suggestive names, such as "maxmin-the-difference" in scores, or "beat-the-average." in certain types of experimental gaming it is important to be aware of these as alternative hypotheses to explain behavior.

Most of the cooperative game solutions presuppose a high level of communication among the players and they presuppose the ability to arrive at agreements and to arrange for sidepayments or some form of *quid pro quo*. The simplest type of cooperative coalitional game uses the characteristic function, as illustrated in Tables 1.1 and 1.2.

A basic underlying assumption about most cooperative solutions, differentiating them from the noncooperative solutions, is that the individuals will get together and play in such a way as to maximize the total benefit to the group, having agreed previously how this

20

benefit will be split among them. Given this assumption, there are still many different ways in which the benefits can be divided. Several different solutions have been suggested and each one of the solutions stresses different conditions for division. The two most widely known solutions are the *core* and the *value*. The core, if it exists at all, is in general not unique. It consists of those sets of outcomes about which no group can complain. If they failed to cooperate, they would receive less by themselves than if they cooperated and were given a payoff within the core.

In the three-person game shown in Figure 1.3 and Table 1.2, any division of 30 is in the core, as no subgroup can guarantee themselves a positive payoff. In the game whose characteristic function is shown in Table 1.3 there is only one point in the core $(1, 1, 1)$

TABLE 1.3

$$v(\theta) = 0$$

$$v(1) = v(2) = v(3) = 0$$

$$v(12) = 2 \quad v(13) = 2 \quad v(23) = 2$$

$$v(123) = 3$$

The value solution is one based upon axioms of symmetry and fair division. It can be looked upon as a suggested fair division procedure. There are other solution concepts such as the stable set, which can be given a sociological interpretation, and the bargaining set, which reflects individual bargaining powers of the players. There are considerably more than a dozen solution concepts of a cooperative variety which have been suggested. Each one picks up a different facet of the social complex of cooperation and competition.

On Problems in Modeling

In the previous section we noted six problems in modeling. Prior to discussing them, a thumbnail description of game theory has been given. We now return to the six problems.

(1) The Definition of Rules and Problems of Wording and Coding

It might be hoped that the theory of games would provide a methodology for the description of all games; indeed for games such as chess, bridge, poker, tactical tank battles, or business games it does provide an extremely powerful methodology. However, in situations involving verbal interchange, body movement, facial expressions, or any other possibly ambiguous, nonnumerical type of communication, the task of specifying rules, strategies and payoffs calls for behavioral insights that are not supplied by game theory. A quick and easy check as to the difficulties of description can be made by asking how easy it is to write a simple computer program to describe these phenomena. For example, it is extremely easy to describe a move in the game of chess in an unambiguous manner to a computer. A phrase such as "He sneered as he replied" cannot be so easily coded.

The methodology of game theory offers no way to cope with general "coding problems." The answer to this problem lies with other disciplines; hence, the social psychologist, psychologist, sociologist, and psychiatrist must supply interpretations that the economist, game theorist, or mathematical modeler cannot hope to supply alone.

Much of the confusion about and some of the misapplication of game theory has been caused by a lack of understanding and appreciation of the fact that the formal theory of games makes no claims to have solved the critical coding problems of how to represent and interpret verbal or other signalling behavior in terms of moves. When the tradeoff is between mathematically analyzing the wrong problem or having a poorly formulated problem but formulating the correct problem even though one's analysis is weak, frequently the latter is the more profitable course.

(2) and (4) The Definition of Rules, the Meaning of Rationality, and Problems of Information and Data Processing Ability

Problems concerning information, rational behavior, and data processing are also closely related to the distinction between environment-rich and environment-poor games, and as such we discuss these two points together. In all of the uses of gaming there has been a constant difference of opinion concerning the importance of levels of complexity, richness of the model, and control over the game. Economists, game theorists, and some political scientists have tended to use a rational model of man who knows what he wants, has well defined goals, and attempts to optimize the attainment of his goals. He is frequently modeled as being completely informed and able to utilize the information he has available. In contrast, a different model of man is that he may not know all of his alternatives, he may not even be certain about the alternatives of which he is aware, and his information-processing capabilities are limited. Give him too much information and it may be as or more damaging than too little.

Even if it were accurate to assume that individuals were capable of evaluating any set of alternatives, when a game becomes sufficiently large and complex, even though the builder may claim that it is completely defined, it may be too large for a single individual to know all of the rules. At that point, the game becomes somewhat free-form in the sense that players discover rules that they have failed to observe previously, and, if things are sufficiently complex, they may need referees and lawyers to give the current interpretation of what the mass of different rules mean. Big, computerized business games and large military games verge on becoming partially free-form due to their complexity. One of the key problems in gaming that is closely related to the formal description of a game is what effect different scenarios have if they are written about a simple game, which, in each case, has the same basic analytic structure. For example, if the same game is presented as a business game, the variables are then redefined and it is dressed up as a military scenario, or as a political problem, will there be any difference in the way the game is played, given the fact that the mathematical structure in each case is the same?

(3) The Specification of Payoffs, Goals, and Motivation

The game theoretic view of the individual is simplistic in the extreme. In general, each player is assumed to try rationally to maximize a utility function or some specified payoff. In actual games, boredom and distraction play roles not accounted for in game theory. Players may not even fully know their preferences and, if money is involved, the stakes may not necessarily be sufficiently interesting. Furthermore, as is noted in Chapter 6, the actual goal of the players in a gaming exercise may have very little to do with their formal stated goals. For example, in a war game, the goal of a colonel "with stars in his eyes" may be quite different with senior personnel looking on, than when he is the senior ranking officer present, even though his role in the game is the same.

(5) The Meaning of Rationality and the Concepts of Solution for Multiperson Games

We have already sketched some of the solutions to a multiplayer game. Even if we were to assume that all individuals were rational, well-informed decision makers with clear individual preferences, one of the most important lessons that game theory has to teach is that a concept of individual rationality does not generalize in any unique or natural way to group or social rationality. Social rationality could easily be a concept defined independently from individual rationality and may not even be consistent with it.

The meaning of social rationality is one of the prime concerns of game theory. The different attempts to define a solution to an n-person, nonconstant sum-game amount to suggesting different criteria for social rationality if they are offered as normative solutions.

(6) The Specification of Players as Individuals or Groups

The atom, or the basic unit in game theory, is the player as the individual decision maker. In gaming, the player may well be an individual, a formally structured group, or an informally structured group. Once a game becomes sufficiently complex, there is a need for a division of labor in decision making and information processing. The basic theory of games has nothing in particular to say *a priori* about the theory of the structure of bureaucracies. It could be developed to provide insight if the assumptions are made that information processing takes time, costs, and resources and that the obtaining of certain types of information and communication is not free.

It is the author's belief that in operational gaming, be it military gaming, urban planning, or political gaming, or the use of large scale business games, not enough attention has been paid to bureaucratic problems.

A basic knowledge of game-theory methods for modeling and analysis can be of considerable use in separating out variables and in isolating problems in the construction of games. A mixture of both game theory and behavioral theory is called for if we are to develop successful ways in which to analyze process models of decision making.

A CONTRASTING OF GAME THEORY AND BEHAVIORAL ASSUMPTIONS

In closing this chapter, a brief contrasting of some of the basic assumptions underlying the models usually employed by game theorists and underlying the models used by social psychologists and other behavioral scientists when experimenting with games is given.

Game Theory	Behavioral Theories
Rules of the game	Laws and customs of society
External symmetry	Personal detail
No social conditioning	Socialization assumed
No role playing	Role playing
Fixed well-defined payoffs	Payoffs difficult to define and subject to change
Perfect intelligence	Limited intelligence
No learning	Learning
No coding problems	Coding problems
Primary static	Primarily dynamic

This list demonstrates the fact that the approaches of the game theorist and the behavioral scientist are complementary. It is not by accident that when they both do an experiment on, say, a game such as the Prisoner's Dilemma, they are looking at extremely different phenomena. For example, one of the most important underlying assumptions in most game-theory models is that of "external symmetry." Unless otherwise specified, it is assumed that all players have the same intelligence, personality, ability to compute, and so forth. Much of the mathematical ingenuity of the game theorist goes toward finding out how symmetrically described individuals will play in lopsided or nonsymmetric games. In contrast, many of the social psychologists' experiments are aimed at finding out how different types of individuals play in symmetric games.

CONCLUDING REMARKS

One brief chapter can scarcely do justice to the description of the various theoretical underpinnings for gaming of any variety. Even though this book is mainly concerned with providing some "how to" information and reference on various types of gaming, it would be remiss to fail to stress the need for a development of a theory of gaming to parallel its practice and to try to sketch the different roles that simulation techniques, game theory, and behavioral theory must play in the successful development of gaming.

REFERENCES

1. Forrester, J.W., *World Dynamics*, Cambridge, Mass.: Wright-Allen, 1971.
2. Shubik, M., *Games for Society, Business and War*, Amsterdam: Elsevier, 1975.

2 On the Many Goals of Gaming

Professional standards in gaming are only beginning to emerge. People who regard themselves as involved in gaming or simulation still tend to use the same words for very different concepts, depending upon their own interests. It is certainly premature to believe that there is such a thing as "the method for evaluating or validating all games." For the question of validation to be answered meaningfully, the sorting out of the different uses of gaming and the development of criteria and methods that apply to these different uses is essential. In this chapter a sketch of the different uses of gaming is given.

If we use the word "validation" to apply to statistical tests concerning the validity of inputs, functional forms, or outputs of a game or simulation, or to experimental comparisons between the simulation and other representations of the phenomenon and the phenomenon itself, then, before we even approach the problems of validation, we must deal with the equally important aspects of intention, specification and control. Intention is discussed here; the others are discussed in Chapter 6.

The promise from many of the different types of gaming appears to be considerable. The proof of the promise is by no means empty for some of these categories. There are some reasonable criteria available for judging the success of a social psychology experiment, and of the teaching value of some elementary games and some business games; the worth of some formal game-theoretic and--simulation models for weapons evaluation; the profitability of sports; and the entertainment value of the theater and entertainment games.

How successful are (and what are the criteria for success of) operational games? What is really learned from political-diplomatic and military exercises? Who learns what from teaching games? The potentialities of gaming are considerable in many different fields of application. The *ad hoc* construction of specification, control, and validation procedures with extreme attention paid to the special purpose at hand could yield valuable insights and results from which the broader generalizations called for by a general theory of gaming might be constructed.

TEACHING

Table 1 shows the six main divisions of the goals of gaming, together with a finer breakdown of these categories.

In teaching and training, the audience for different games is extremely varied with respect to age, occupation, and reasons for using a game. A useful breakdown which correlates well (but not perfectly) with age is the level of education: Preschool, elementary school, high school, undergraduate college, graduate, and adult educational programs.

An individual's occupation and his reasons for participating in a game are highly correlated. Without going into great detail, four reasons are suggested that broadly describe why most players are involved with teaching or training games: (1) They volunteer to play, (2) they are advised to play and follow the advice, (3) they are ordered to play by a superior, or (4) bureaucratic or organizational rules require that they play.

Most games other than sports, in most educational institutions are parts of courses or special programs. Although participation is voluntary prior to registering for the program, once a student is in a program the organizational rules often require that he participate.

Where the participants are members of large bureaucratic organizations, such as the military, the government, or private corporations, they are frequently advised or ordered to participate. Even when they are volunteers, the type of volunteering may be of the type where a department head is told to supply three out of his twenty men for a game. It is not uncommon for the "volunteers" to be the three least busy or most junior men in the department.

A crude estimate indicates that in 1970 there were between 15,000 to 25,000 war-gaming amateurs in the United States[1]. The overall trend in volunteer gaming in the last 30 years has been from an almost exclusive emphasis on military games to military-diplomatic games and to business games, and now more recently to

TABLE 2.1 Purposes of Gaming

TEACHING	EXPERIMENTATION	ENTERTAINMENT	THERAPY AND DIAGNOSIS	OPERATIONS	TRAINING
Motivational aid to learning	Validation of hypotheses	Theater	Group therapy and t-groups	Cross check and validation for other methods	Teaching skills to individuals
Reinforcement for other methods of training	Artificial intelligence	Gambling	Individual therapy	Extra-organizational communication	Teaching bureaucratic organizational behavior
Device for teaching facts	Exploration and generation of hypotheses	Spectator games	Diagnosis	Exploration and testing	Dress rehearsals and "shakedown" exercises
Device for teaching theory		Participant games		Planning	
Device for studying dynamic cases		Solitary games		Group Opinion Formation "Delphi"	
Device for teaching interpersonal relations				Brainstorming	
Enculturation of the child				Forecasting	
				Advocacy	
				"Shakedown"	

28

games concerning society. Currently there is a trend toward games stressing social interaction and the problems of society. This is manifested in the growth of a number of board games that are educational as well as entertaining. Thus we have had a progression from Monopoly to Smog. Even in war games there appears to have been an upswing in the last ten years of games calling for diplomacy, negotiations, and grand strategy, such as Diplomacy and Summit in contrast to straight war games. From a technical game-theoretic point of view there has been a shift from two-person, zero-sum games or noncooperative, individualistic enterprise games to nonconstant sum games where coalitions are of importance.

On Different Roles in Gaming For Teaching

Before questions concerning validation can be asked with respect to a single game, it is desirable to consider goals and criteria of success from several different points of view. In particular, any game should be considered in the context of its impact on the players, the builders, the controllers or directors, and the sponsors. Frequently an individual may play more than one role. Furthermore, the roles are often more finely differentiated than noted above. For instance, the direction of the game may consist of a team containing not only umpires or teachers, but also experts who are called upon to judge the feasibility of certain acts while otherwise having no control role.

At the university level, especially with graduate students, more may be learned by the students in constructing games than in playing them. The learning experience is by no means confined to the players. In gaming used for teaching purposes, especially at or below the high school level, the worth of a game is frequently no more than that of the teacher. An inspired teacher can direct a mediocre game with good results, and the best of teaching games can be ineffective if directed by an inadequate teacher. This breakdown of roles applies also to gaming used for purposes other than teaching, and is referred to again later.

Motivational Aid to Learning

One of the major attractions of gaming has been as a motivational device. It appears to attract the attention of and involve the players deeply, where other methods may have far less impact. There is reasonable consensus on this point among those who have used games, and there is a small acount of experimental evidence as shown by the work of Wing[2] and Boocock and Schild.[3] Creators of educational games, such as Allen,[4] stress the positive motivational features of educational games. However, it is easy to move from conjecture to advocacy as is exemplified by the writings of Abt.[5] Coleman has stressed the value of games in teaching disadvantaged children.[6]

Reinforcement for Other Methods of Training

On all educational levels, games are frequently used to supplement more traditional methods of teaching. This is also true in business schools and military academies. Gaming proponents claim that the mix of methods is most effective.

A Device for Teaching Facts

In virtually every type of gaming, including the diplomatic-military games of the Studies, Analysis and Gaming Agency (the successor to the Joint War Gaming Agency), and business games, such as the Carnegie Tech Management Game[7] or Intop,[8] gaming practitioners and players have claimed that gaming is an extremely useful way to learn and organize facts. A game usually provides a handy scheme for supplying associative links between facts, and as such it may aid both learning and remembering, although to date there is little evidence to substantiate these claims.

A DEVICE FOR TEACHING THEORY

At the advanced undergraduate and graduate level the building of games appears to be extremely useful in encouraging students to think in terms of models and abstractions. This improves their ability to theorize. In the social sciences especially, the importance of improving the ability of an individual to enable him to construct abstract representations of complex systems cannot be overemphasized. The discipline of constructing a playable game provides a deep appreciation of logical consistency and completeness, as well as stressing the connection between the model and its subject matter.

On the other hand, it is important to stress that before a game can be used to teach theory with any success, it must be recognized that the theory exists to be taught. In the exploitation of business games over the last decade this has not always been the case. A flagrant example of potential misuse has been in the modeling of advertising in business games. Even a brief glance at the literature on how advertising affects sales is sufficient to indicate that there is little substantiated theory on advertising, yet in many of the business games played both at universities and in business training programs, advertising has been thrown in as an *ad hoc* modification on demand with teaching results which could be damaging were it not for the basic skepticism of most of the players. It is important that players be warned against learning false or unsubstantiated principles.

A Device for Studying Dynamic Cases

Several business schools, especially the Harvard Business School, favor the use of the case method. A specific historical case may be taken up, a "scenario" written describing it, and the class be required to consider the problems it poses and the solutions to these problems be analyzed.

A game lends itself with great ease to providing a dynamic context to a case. Furthermore, like the Czech experimental theater at Expo '67, it provides a natural means whereby alternative histories can develop. A formal game, especially a large and complex one, has both the advantages and disadvantages of an institution. It may take on the inertia of an institution itself, as is exemplified by the Carnegie Tech Management Game.[9] However, this may be an advantage as it is extremely difficult to explain or reproduce in the classroom the ambience of decision-making within a bureaucracy.

A Device for Teaching Interpersonal Relations

Many of the basic educational games for younger children and disadvantaged groups and as community action games to study urban redevelopment or other social problems stress interpersonal relations from the viewpoint of both the individuals and their roles. In these uses of gaming, seeing the other individual's point of view by role-playing his position appears to be of value. Thus, for example, a slum child may begin to appreciate the difficulties of being a policeman.

At the more direct level an appreciation of the need for communication, bargaining, and compromise can be obtained from many of these games. A good example of such a game is Democracy.[10] Some of the insights gained here do not pertain only to personality factors but to a basic game-theoretic phenomenon that in an *n*-person nonconstant-sum game there is no neat, unique way of defining socially rational behavior." As in society there are many different criteria for social rationality, and it is frequently not possible to satisfy the demands of all groups even if each group can show that its demand is within the scope of its own power if it fails to cooperate with the remainder of society.

Enculturation of the Child

The extremely important role of games in basic learning processes of the child has been stressed by Piaget.[12] There is also a linkage between many games and ancient religious and initiation ceremonies.[13] Although originally there was little formal structure and no direct intent in the development of basic folk games for preschool children, the current interest in learning processes, artificial intelligence, and teaching methods have made the understanding of the role of games at the preschool level an important problem for basic research.

TRAINING

Teaching blends into training, and training blends into operational uses. Nevertheless it is useful to make distinctions among different goals for gaming, although they may overlap. In particular, the major distinction between teaching and training concerns the emphasis placed on the reasons behind the process. For example, there are several quite effective small games which can be of use in improving an individual's performance in production and inventory scheduling without ever going into why certain methods work. An operator does not have to know dynamic or integer programming to become a better manager of production and inventory scheduling.

Many individuals can be taught to drive safely by means of analogue device trainers without having to learn much about Newtonion mechanics or how an automobile works. Training games for simple manual skills, especially those requiring a fair amount of coordination, are not particularly exciting, but they can be of tremendous use and can provide valuable simulated experience that would be extremely costly to obtain from the field.

In general, when games are used for training, the only role occupied by the individual being trained is that of player. This contrasts with gaming for teaching, which, because the *why* is so important, it is highly desirable in some instances to have students build or supervise as well as play games.

Bureaucratic and Organizational Behavior

In a complex society, licenses must be obtained, permits granted, rules checked, exceptions examined, accounts audited, telephone calls made, and routines for processing torrents of communication must be established. Training games offer the possibility not only for training individuals to acquire individual skills but also to learn bureaucratic routines.

Dress Rehearsals and Shakedown Exercises

Rehearsals in the theater, field maneuvers, and battle exercises are all examples of operations devoted to seeing that individuals know their own roles and are able to cooperate in team action. They differ from the previous category only inasmuch as they are usually aimed at preparing for coordination in a temporary context, such as a specific play or a projected offensive. The word "shakedown" appears to come from the naval usage "shakedown cruise" which is the original cruise of a ship devoted to coordinating the crew and checking the equipment.

OPERATIONAL GAMING

In contrast with gaming for teaching, operational gaming is used almost exclusively by adults in military, government, or corporate organizations. There is an overlap between operational and training games in the domain of field exercises. It is difficult to say where the dress-rehearsal and coordination aspects of an exercise cease and where planning, strategy, testing, and exploration begin. The category "shakedown" is as relevant to operational gaming as it is to training. By far the largest use of operational games to this day is in the military or diplomatic military. Relative to these uses, corporate operational gaming is insignificant and the use of operational gaming for social planning is still in its infancy.

Because of the nature of the bureaucratic structure surrounding decision making, a clear understanding of the roles and goals of the players, builders, controllers, and sponsors of operational gaming exercises is far more important to the professional who wants to know what is going on than is such detailed understanding of the use of gaming for teaching.

Operational gaming is "where the money is" currently, and the goals of a consulting firm wanting to build a large game, a general wishing to advocate a weapons system and a colonel assigned to play in or operate the game can be sufficiently diverse that the mismatch makes an objective evaluation of such a game impossible.

Cross-Checking and Extra Validation for Other Methods

A game may be used as a back-up procedure to provide an extra insight into a process that has been investigated by other means. For example, a recommendation may be presented in report form, the basis of which may be expert opinion and/or empirical evidence. A gaming study of the same problem may turn up insights or raise questions overlooked by the report. As operational games in general tend to be somewhat expensive in both time and money, the problem has to be of sufficient importance to merit the extra effort. There is also the danger that a game may be employed to give a pseudoscientific endorsement to a recommendation.

Extraorganizational Communication

There may be a game outside of the game being played. With operational games it is critical to understand both the stated and the unstated purposes of gaming by the individuals involved in the exercise. In particular, gaming, along with short courses and seminars, is used to establish informal means of communication. In some instances the main objective may be to arrange to get two or three ranking individuals trapped together for two or three days on neutral ground. Participants in diplomatic-military war games frequently comment on the value of being able to watch the decision-making styles of different high-ranking individuals.

The use of a game as a means for establishing informal communication will vary heavily with the style of play. If the game is held in an isolated locale over an intense period of play for three of four days or more, the effect may be quite striking. If, on the other hand, it is played in an intermittent manner over several weeks or months, then it is easy for most of the participants to minimize the disturbance to their set patterns.

Exploration, Testing, and Planning

The strict meaning of a strategy in the sense of game theory, while precise and worthy of note to a gamer, is not particularly useful to a planner. Planning involves the selection and aggregation of information. Even with the aid of high speed digital computers the number of alternatives which can be explored is miniscule. Games such as those played by the SAGA[14] operation or the Sierra series[15] of the Rand Corporation and many others have been used for planning, exploration, and the testing of a limited number of alternatives.

An intense amount of preparation goes into a game of this type. The preparation is in general far more extensive than the play. Two or three moves on each side may be taken, and in a debriefing session after the game there will be an attempt to summarize and note the consequences, alternatives or facts that had been overlooked before starting.

A planning game to be of use must utilize individuals sufficiently involved in the process that they can be privy to the actual problem and the major considerations. In military and governmental games these may range from colonel to five-star generals and cabinet officers. There is some evidence that some high-ranking officials enjoy participating in gaming exercises; but there appears to be little actual evidence as to what was accomplished. This last comment applies to gaming regarded as a "brainstorming" exercise as well.

Group Opinion Formation and Delphi

In the behavioral sciences and in the study of organizations, in evaluating the present and in forecasting the future, we have very little documented knowledge in the same sense as in the sciences in which experiments are performed and *replicated* frequently. Much use is made of expert opinion, but until recently little systematic thought has been given to the study of how expert opinion is used and what the techniques are for optimizing its use. Furthermore, little emphasis was placed on the relative worth of using the opinion of more than one expert. For example, when do diminishing returns set in and what sort of controls should there be over the interaction?

An operational game may be regarded as a formal structure to elicit group planning, a process involving both evaluation and prediction of contingencies.

Helmer[16] and Dalkey[17] have advocated the use of Delphi techniques, which consist of having a group of experts who are anonymous to each other respond to questionnaires, after which the results of their responses are processed and returned to them so that they can benefit from the views of their colleagues. Dalkey is engaged in large-scale experimentation[18] on the properties of the Delphi method.

One important feature that differentiates a formal operational game from Delphi is that a formal operational game has more emphasis on the aspects of motivation in relation to performance than Delphi. To date there has been little effort to blend these two approaches. However, the potential appears to be worthwhile.

Forecasting

In general, a game is *not* a forecasting device. A good operational game may make use of good forecasting procedures but it is not in itself aimed at providing forecasts. This should not be confused with its use in discovering new alternatives and in helping to evaluate future possibilities. Forecasting and contingency planning are related but extremely different activities. In particular, a good forecaster is not necessarily interested in the importance or worth of his forecast. Accuracy may be a goal for the forecaster in and of itself, not because of its relevance to the planning process.

A game may be a useful device for stressing the need for coordination of forecasting activities with planning and decision-making processes. In this sense the involvement of forecasters in the design and play of operational games may be of considerable use.

Advocacy

Last, but not least, we must note the use of operational games for advocacy. A competent game designer can build in biases into a game. Advocates for specific policies or weapons systems can load the dice so that the game has a great probability of producing the results they want to see. Games are fun; they are also great propaganda devices. The AMA business game provides one such example.[19]

Action groups of nonprofessionals can easily be hornswoggled by a latterday snakeoil salesman peddling a game to cure all ills. Smog, fog, the crime rate, central city decay, impotence, war, lack of understanding among nations, the evils of unemployment and the drug culture, the curse of the automobile, and the lack of a good 5¢ cigar will all be cured if we only have a big enough data bank tied into a game room with large, fancy maps.

Recently there has been a move for the building of a "World Game" by several extremely well-meaning individuals.[20] As a mild advocate of gaming this author believes that there are many good reasons to proceed with the use and building of large games for operational purposes, especially in areas dealing with social policy. However, one must not confuse conversational feasibility with operational feasibility.

In some instances a game can be used as a euphemistic way for informing others of a change in policy by asking them to participate in an exercise whose outcome is a foregone conclusion. The Japanese war gaming prior to Pearl Harbor could be interpreted in this manner.[21]

EXPERIMENTAL GAMING

Human beings fortunately are more difficult to experiment with than rats or guinea pigs. Even so, there is now a fast-growing body of literature on experimental gaming in which human decision-making behavior is studied by observing the performance of individuals in formally structured games. In order to pursue this type of work fruitfully it is important that the experimenters have at least a basic elementary understanding of game theory and social psychology.

Much experimentation has been done with simple 2×2 matrix games under relatively restricted conditions. The experimental subjects have been, for the most part, undergraduates at various universities; some army personnel have been used, as have some middle- and upper-level corporate personnel.

These experiments are very different from preschool educational games or from military-diplomatic operational free-form war games. The criteria for validation belong to more-or-less accepted statistical methods familiar to physical scientists, econometricians, and experimental psychologists.

Some experimentation has been performed with business games of moderate or considerable complexity[22] and with political, diplomatic, and war games.[23] In general, due to the greater complexity and smaller degree of control on these games they have been harder to control, and hypotheses have been difficult to test. In some instances (Hoggatt,[24] Shubik, Wolf, and Lockhart[25]) players have been faced with artificial players as competitors.

Validation of Hypotheses

In general, although the goals of the game designers are usually clear in experimental gaming, the goals of the players are by no means clear. There exists an enormous and frequently poorly handled problem in specifying, controlling, and measuring the goals and motivations of players in simple as well as in complex experimental games.

A separate chapter would be needed to deal with the literature on experiments with 2 X 2 matrix games, and another needed to discuss experimental work on the analysis of human factors in complex competitive systems. Nevertheless, several disturbing features of work in gaming should be mentioned. Specifically, much of the work with operational games presupposes that a considerable number of problems that belong in the domain of experimental gaming, or basic research, have been solved, whereas, in fact, the expenditures and activities in experimental gaming are miniscule as compared with operational gaming.

Furthermore, although the word "validation" is popular and takes on a particularly scientific flavor when applied to experimental games, if one does not know what one is trying to validate, then all of the statistical apparatus available may be useless. *Control* and *specification* are prevalidation conditions that are not yet carried out adequately on many of the experiments. The major contributing factors to the failure of control and specification are lack of cooperation among specialists (for example social psychologists who know no game theory misunderstanding the competitive structure, or game theorists knowing no social psychology failing to allow for simple explanations of behavior), and lack of sufficiently automated laboratory facilities to enable the careful experimenter to obtain detailed observations and to run standard analyses at a reasonable cost.

Artificial Intelligence

In the past decade there has been a considerable upsurge in the study of artificial intelligence, or in the study and the construction of computer programs that perform tasks that are usually regarded as requiring intelligence. No distinction has been made, in general, between the sort of intelligence required to solve difficult problems, such as playing chess, and to resolve interpersonal problems, such as those which arise in nonconstant sum games, such as bargaining. Frequently the gamer is more interested in social intelligence than in individual intelligence. The problems in the construction of a good problem-solver or a socially intelligent player differ inasmuch as the criteria for the performance of the former are relatively easy to construct, whereas there are no such easy criteria that can be constructed to judge group or social rationality. In particular, it appears that a good problem-solver—a program which can play chess well, for instance—requires efficient searching and calculating abilities and other features usually associated with intelligence and intellect. By the very nature of the game it need not have any "personality." A good chess-playing program has to be an intelligent program, not a pleasant or nice one. This is not the case when we turn to nonconstant sum games. It is possible to build an artificial player for a business game[26] which plays in a manner comparable to human players. The rules or heuristics needed to construct such a player call more for an emphasis on his interpersonal relationships than on his ability to compute. A "nice," moderately cooperative, and not particularly aggressive artificial player in a business game may elicit cooperation from his competition and will do quite well.

The literature on artificial intelligence has very little on the subject of social intelligence. There has been and is currently an extreme division of opinion on the nature of problem solving, leaving aside the extension to social interaction. Simon, Minsky, Papert,[27] and many others are the proponents, whereas considerable criticism of the basis of artificial intelligence work has been offered by Bar Hillel and H. Dreyfus.[28]

Along with the growth of interest in artificial intelligence has come a considerable growth in the design of protocols and ways to describe decision-making processes. Much work has been addressed to analogies between how one teaches a machine and how one teaches a child.[29] In particular, those interested in experimentation with computer-aided instruction[30] need to be aware of the developments in artificial intelligence.

The experimental gamer is usually more interested in games that are more than problem-solving exercises. Many war games and games such as chess can be modeled as two-person zero-sum games. Hence the main analytical problems they pose are in the domain of information processing and problem solving. Diplomatic-military, business, social-development, and most other games do not fall under the zero-sum rubric. Social, political, or economic behavior all call for attention to interpersonal interaction. The construction of robots or artificial players in these games both provides opportunities to attempt to model sociopsychological processes in the building of the players and gives the experimenter greater control over his experiments, especially when he is able to replace a set of two-person experiments with a set of experiments consisting of a group of individual human players playing with the same artificial competitor.

Exploration and Generation of Hypotheses

Frequently, experimental games are used to explore decision-making processes and to generate new hypotheses rather than to test existing hypotheses. Prior to an experiment, several hypotheses may be set forth, but after the experiment it appears that these hypotheses can neither be accepted nor rejected, owing to insufficient definition or complications in the control of the experiment. Nevertheless, the running of the experiment clarifies the definition of the hypotheses, locates others, and locates the control difficulties.

The above reasoning is often used as an excuse or self-justification after an ill-conceived experiment has been run. However, this is not always the case and pilot experiments play an extremely useful role when the topic being studied is both complex and ill-defined.

GAMES FOR ENTERTAINMENT

The Theater

It is important to remember the deep interconnection between gaming and theater. For example, many war exercises, fleet maneuvers, and "dry runs" are identical in purpose with dress rehearsals. Huizinga[31] and Callois[32] are among those who have discussed the relationship between plays and games. It is not the purpose of this volume to explore the historical, anthropological, and religious aspects of this interconnection. However, those who wish to use games for more mundane purposes should at least be aware of the interrelationship among games, plays, theater in general, mass spectacles, such as circuses, coronations, public executions, and ceremonial parades. For example, a ceremonial military parade itself is an extremely complex phenomenon, being part entertainment, part training, partly a signaling process in a diplomatic dialogue, and a device for influencing morale.

An important but open question is what the basic features are that differentiate good theater from good operational gaming. For example, how does the "realism" of the scenery affect both of these activities? The audiences are different, the role-playing is different, and the stated purposes are different. Nevertheless, an analytical categorization of these differences is not an easy task.

Gambling

Three categories of individuals involved in gambling must be distinguished. They are those businessmen who run gambling ventures, professional gamblers who make a living from playing, and those who play for other reasons.

The individuals who run gambling establishments are in many senses not particularly distinguished from other businessmen except that possibly gambling as a business tends to be an enterprise with not very large components of risk as compared with a highly innovative technology enterprise. The professional gambler, such as the poker player (see, for example, Yardley[33]) does not seem to be far different from the professional arbitrager. They both take risks but they are in the true sense of the word calculated risks and the individuals who devote professional attention to these occupations are usually skillful enough that they are able to earn a good (but, in general, not spectacular) living from their professional skills.

Although there is a large element of chance in a game such as poker, in contrast with a game such as roulette, poker is primarily a game of skill and not of chance. The calculation of probabilities is one of the key aspects of good poker playing. There are obviously enormously psychological factors important in one's ability to judge the competence and style of the other players. Betting on horse races stands somewhere between a roulette game and a poker game in its skill component. The element of chance is extremely large. However, there are some useful calculations to be made about the odds being offered and the probable performance of the horses. The main factor, however, does come in the judging of the horses and their performance on a specific track under the appropriate weather conditions.

In general, especially in large organizations, when someone states "we have taken a calculated risk," that frequently means that individuals have made a decision without doing the calculations necessary. In the case of the professional gambler, the reverse is true. In general, they have no bureaucratic structure around them, and fast and explicit calculation of the odds is a central aspect of their very living.

In contrast with the fighter pilot, the poker player thinks explicitly in terms of odds. It is unlikely that the pilot calculates a probability of .15 of success by one avenue or .3 of success by another approach. There is undoubtedly an important psychological difference in calculating explicit probabilities involving death and explicit probabilities involving money. Furthermore, the nature of many gambling games of skill makes the calculation of probabilities a natural and explicit way of evaluating one's position. It is unlikely, however, that these features are in themselves sufficient to explain the fundamental difference in approach to thinking in terms of explicit probabilities evinced by professional gamblers as compared with, say, middle management men, army generals, or, even more so, the citizen on the street. The literature on the social and personality aspects of the professional gambler appears to this author to be surprisingly slight. This also holds true for the handful of special professions in which the risk of life component is sufficiently important to make the gambling aspect explicit, for example, test pilots, stunt men and steeplejacks.

The interests of the individuals not professionally involved in gambling run the gambit from mild entertainment to deep addiction. Many individuals who lose $20, $30, or $100 at the tables in Las Vegas or Monte Carlo are paying an entertainment fee. For the most part, they know that they are paying this fee and have decided that the entertainment is worth it. It is worth noting that the mere location and decor of major gambling towns and main casinos lay stress on the theatrical aspects and the role-playing features of gambling. Las Vegas is designed so that the perfectly ordinary middle-class dweller of suburbia can lose his $100 or so in a socially acceptable manner in surroundings with components of pseudoluxury and pseudowickedness.

What are the risk-taking characteristics of the ordinary individual who is not addicted to gambling, who plays small-stake roulette at a casino, or who buys an occasional ticket for the races? There exists a certain amount of literature in economics and psychology concerning gambling and the buying of insurance where the odds are in general extremely small for an event to occur. However, there is very little analytical literature on gambling behavior. Goffman has several highly stimulating articles on con-games where the otherwise prudent and nonaddicted individual is "taken for a sucker."[34]

There are many individuals for whom gambling is an addiction. Dostoevsky was a good example of one of these. There is some psycho-pathological literature on gambling an example of which is by Bergler.[35] One of the difficulties in studying a subject such as pathological gambling is that it requires a multidisciplinary approach. Psychiatrists will tend to see only the psychological aspects whereas, for example, those trained in a theory of games will undoubtedly lay heavy emphasis on the structural differences among various games.

From the viewpoint of those interested in operational games, especially games of a military or social variety, the study of addiction and extreme risk-taking would appear to be critical. The distance between the gambling addict and the drug addict may not be great. There also appears to be an important psycho-pathological risk-taking component in assassinations and in the actions of some extremist groups.

In summary, it appears that gambling behavior of virtually all types is a critical phenomenon in the understanding of many important features of risk-taking. Those who argue for operational games as a means for studying extremely original or surprising alternatives should also consider the need to explore the genesis of both "reasonable" and pathological risk behavior.

Spectator Games

Many sports, such as football, baseball, hockey, basketball, and soccer are primarily spectator sports. The vast majority of the participants are in the audience and derive vicarious pleasure from the play. There the analogy between the game and theater is possibly at its closest. The great majority of participants are spectators. The sports event is far more of a free-form play than is a theatrical performance. In the former, although the rules are set the actual path of the play is not completely known in advance. In the latter, the complete path of the play has been specified except for what small portion of the acting that has not been controlled by the director. Spectator games may have a small advocacy and teaching component to them, inasmuch as they may inculcate an appreciation of teamwork and an ability to judge and understand the qualities of effective performance. However, for the most part, they are pure entertainment. For a discussion of the vicarious pleasure and role identification aspects of spectator entertainment, see Callois.[36]

Participant Games

Bridge, poker, tennis, chess, golf, charades, Monopoly, and many board games, many of which can be played as spectator games, are most frequently played only by the active participants for their own amusement.

A controlling factor in determining whether a poker game is played for amusement or for gambling purposes is the size of the stakes. The importance of the payoffs to the players as an influence on the nature of the game cannot be stressed enough. When an individual participates in a game whose stated purpose is operational or educational, but which nevertheless is formulated in such a way that the payoffs to him are not particularly clear, it becomes absolutely crucial to investigate the possibility that he has turned the exercise into a game for his entertainment. It is safe to have as a working hypothesis when using games for teaching, experimentation, operations, or therapy, that in fact the game was primarily theater or participant entertainment until proved otherwise.

Solitary Games

Possibly one of the greatest sinks for the use of man hours in gaming is the solitary game. Crossword puzzles, jigsaw puzzles, and solitaire are major examples of games played to pass the time, although it can be argued that they may have an educational component. The origins of both the crossword puzzle and the jigsaw puzzle are relatively recent (within the last hundred years). Precisely what makes them so popular? Will they be supplanted by other solitary games? Could solitary games be designed that would be fun and more explicitly educational or experimental?

THERAPY AND DIAGNOSIS

Games have both diagnostic and therapeutic value. Although these areas lie well beyond the scope of this book, a useful bibliography is given in Avedon and Sutton-Smith.[37]

Group Therapy and T-Groups

In some ways group therapy sessions and T-groups might be regarded as "anti-games"; as such, the comparison between them and a formal operational game, such as a diplomatic-military game, becomes of considerable interest. In an operational game the individuals are encouraged to concentrate on certain aspects of role-playing. Frequently an individual is required to simulate the decision-making process of someone else. In contrast with this activity, in group therapy individuals participate in a discussion of their problems in an attempt to solve them. The stress appears to be away from role-playing. It appears that the paradigm of the game offers an extremely fruitful basis for joint work by psychiatrists, social psychologists, and those interested in organizational decision making.

It is not difficult to design games that focus on relatively narrow band-widths of decision-making and of interactive behavior. Informal experimentation with several games, such as So Long Sucker[38] and The Dollar Auction,[39] indicate that it is possible to obtain extremely strong participant reactions to relatively simple games. Experimentation with 2 × 2 matrix games of certain design has also indicated this. The use of small games for diagnosis might be relatively cheap and effective. The use of games for therapeutic and corrective purposes is clearly closely related to, but somewhat different from the use of games in teaching. This author knows little about the potentialities of this use, hence refrains from further discussion here.

CONCLUDING REMARKS

The scope of gaming is considerable. Its uses are varied both in concept and purpose. Yet at the same time amid all of the diversity a certain common thread is present. The game is a paradigm for competitive and/or cooperative behavior within a structure of rules. The rules vary in formality in free form gaming or in rigid-rule gaming, and they vary in portrayal of war situations, economics, and social contract formation. But *all* games call for an explicit consideration of the role of the rules. A serious user of games is well advised to be broadly aware of the alternative uses and meanings of games as well as deeply specialized in his own type of gaming.

The prime purposes of the classification given above are: (1) to call attention to the important prevalidation problems of intention, that is, stating purpose and devising criteria by which to judge the attainment of one's goals; (2) to indicate the possibility that *in spite* of the diversity there may be a common core of knowledge and professional skills of importance to all gamers; and (3) to suggest that all specialists stand to benefit from an understanding of the diversity of gaming because frequently different types of gaming overlap and errors or important phenomena that may be completely ignored by one specialist may be obvious to another who sees the same game from a somewhat different viewpoint. (A different classification has been suggested by Bowen[40]).

EXERCISES

1. Contrast the differences between theater and operational gaming in terms of scenarios, role playing, and other features which you deem can be contrasted.
2. Discuss why you believe that it is or is not relatively easy to build an artificial player for a (nonconstant sum) business game which performs as well (measured in terms of profit) as a live player.
3. Discuss why a good chess-playing program should be easier or harder to build than a program to simulate a player in a business game.
4. For what gaming purposes do you believe that a game simulating election processes would be useful?
5. Specify what is meant by the term "calculated risk."
6. In terms of skill and chance contrast roulette, poker, and betting on horse races.
7. Do you believe that there is any difference in the way individuals evaluate the odds to events involving death as a possible outcome and in the way they evaluate risk when only money stakes are involved? Discuss the reasons for your beliefs and consider mountaineering, exploring, sportscar racing, buying lottery tickets, and buying bonds or common stock.

REFERENCES

1. Based on the number of subscribers to the Avalon Hill publication, *The General*. (See also *Tactics*).
2. Wing, R.L., *The Production and Evaluation of Three Computer-Based Economics Games for the Sixth Grade*, Westchester County, Board of Cooperative Educational Services, 1967.
3. Boocock, S.S. and Schild, E.O., *Simulation Games in Learning*, Beverly Hills: Sage Publications, 1968.

4. Allen, L.E., *Wff'n Proof: The Game of Modern Logic.* Chicago: Science Research Association, 1963.

5. Abt, C.C., *Serious Games,* New York: Viking, 1970.

6. Coleman, J., "Social Processes and Social Simulation Games," in Boocock, S.S. and Schild, E.O., *Op. Cit.* 29–52.

7. Cohen, K.J., Dill, W.R., Kuehn, A.A. and Winters, P.R., *The Carnegie Tech Management Game,* Homewood, Ill.: Richard D. Irwin, 1964.

8. Thorelli, H. B. and Graves, R.L., *International Operations Simulation.* New York: Free Press, 1964.

9. Cohen, K.J., Dill, W.R., Kuehn, A.A. and Winters, P.R., *Op. Cit.*

10. Coleman, J., *Democracy* Baltimore: The Johns Hopkins University, Department of Social Relations and Academic Games Associates.

11. Shubik, M.. "On Gaming and Game Theory," *Management Science,* 18, 5. January 1972 37–53; and Shapley, L.S., and Shubik, M., *Competition, Welfare and the Theory of Games,* (manuscript in process), Chapter 6.

12. Piaget, J., *Play, Dreams and Imitation in Childhood,* Trans. by C. Gattegno and F.M. Hodgson, New York: Norton, 1962.

13. Sutton-Smith, B., Roberts, J.M., and Kozelka, R.M., "Game Involvement in Adults," *Journal Sociology Psychology,* 1963, 60, 15–30.

14. The type of game used here was originally suggeste by Goldhamer. See: Goldhamer, H. and Speier, H. "Some Observations on Political Gaming," *World Politics,* 12 (1959), 71–83.

15. See,Northrop, G.M., *Use of Multiple On-Line, Time-Shared Computer Consoles in Simulation and Gaming,* The Rand Corporation, Santa Monica, P-3606, 1967.

16. Helmer, O., *A Use of Simulation for the Study of Future Values,* Santa Monica, Calif: Rand Corporation, P-3443, 1966.

17. Dalkey, N.C., *The Delphi Method: An Experimental Study of Group Opinion,* Santa Monica, The Rand Corporation, RM-5888-PR, June 1969.

18. Dalkey, N.C., and Roarke, D.L., *Experimental Assessment of Delphi Procedures with Group Value Judgments,* The Rand Corporation, R-612-ARPA, Santa Monica, Calif.: February 1971.

19. Bellman, R., Clark, C.E., Malcolm, D.G., Craft, C.J. and Ricciardi, F.M., "On the Construction of a Multistage, Multiperson Business Game," Journal of the Operations Research Society of America, V. 5., 4, (August 1957).

20. Fuller, B., *Presentations to Congress: The World Game,* Carbondale, Southern Illinois University, 1970.

21. See Wohlstetter, R., *Pearl Harbor: Warning and Decision,* Stanford: Stanford University Press, 1962.

22. Hoggatt, A.C., "Measuring the Cooperativeness of Behavior in Quantity Variation Duopoly Games," *Behavioral Science,* 12, 2 (March 1967.) 109–121.

23. Hermann, C.F., "Validation Problems in Games and Simulations with Special Reference to Models of International Politics," *Behavioral Science,* 12 (May 1967), 216–231.

24. Hoggatt, A.C., *op. cit.*

25. Shubik, M., Wolf, G. and Lockhart, S., "An Artificial Player for a Business Market Game," *Simulation and Games,* 2, (March 1971), 27–43.

26. Hoggatt, A.C., *op. cit.*

47

27. Minsky, M., and Papert, S., *Artificial Intelligence Memo No. 200, Progress Report 1968–69*, Cambridge Mass., 1970.

28. Dreyfus, H., *What Computers Can Do: A Critique of Artificial Reason*, New York: Harper & Row, 1972.

29. Minsky, M., and Papert, S., *op. cit.*

30. See, Atkinson, R.C., "Role of the Computer in Teaching Initial Reading," *Childhood Education*, 1968.

31. Huizinga, J., *Homo Ludens*, Beacon Press, Boston, 1955 (translation).

32. Callois, R., *Man Play and Games*, London; Thomas & Hudson, 1962.

33. Yardley, H.O., *The Education of a Poker Player*, New York: Simon & Schuster, 1957.

34. Goffman, E., "On Face Work," *Journal for the Study of Interpersonal Processes*, 18, 3 (August 1955), 213–231.

35. Bergler, E., *The Psychology of Gambling*, New York: Hill & Wang, 1957.

36. Callois, R., *op cit.*

37. Avedon, E.M., and Sutton–Smith, B., *The Study of Games*, New York: Wiley, 1971, Chapter 8.

38. "So Long Sucker," A Four-Person Game, 1951, M. Hausner, J.F. Nash, L.S. Shapley, and M. Shubik, in Shubik M, Game Theory and Related Approaches to Social Behavior, New York: Wiley, 1964, pp. 359–361.

39. Shubik, M., The Dollar Auction Game: A Paradox Noncooperative Behavior and Escalation,: *The Journal of Conflict Resolution*, 157, 1 (1971), 109–111.

40. Bowen, K.G., *The Structure and Classification of Operational Research Games*. Washington, D.C.: Defense Operational Analysis Establishment, Memorandum 7117, June, 1971.

3 Techniques, Modeling, and Languages

DATA PROCESSING AND SIMULATION

In the construction of a complex game or simulation, analytical model building, econometric and sociometric methods, together with data gathering and data-bank organization, are complementary approaches. Gaming and simulation place emphasis upon control (as in a feedback process). Econometrics and the gathering of data and data-bank construction are more concerned with measurement and forecasting.

Frequently the question is asked: Is a game useful for forecasting? In general the answer is no. It is no better for forecasting than the data gathering, modeling, and econometrics that went to make up the model of the environment. A game or simulation is best used for planning, managing, and anticipating. A simple example may help to illustrate this point. A problem in econometrics might be to estimate the number of automobiles sold next year, on the assumption that certain conditions in the environment can be taken as given. A problem in anticipation and control is to estimate the expected worth (according to some criteria) of performing and act that might influence the number of automobiles sold. The first serves as a natural data-input for the second, but not vice-versa.

The problems of control, planning, and forecasting are highly related but are nevertheless different. The statistician and econometrician and other data processors are less concerned with the cost and value of their investigations in terms of policy than they are with meeting the various criteria of accuracy. In general they are not in a position of control and thus it is most reasonable to ignore the effects of interaction between themselves and the system. However, when we consider those interested in using games to study urban development or using operational games to study international affairs, it may be critical to take interactions among the players into account. Furthermore, the problems that must be coped with are not merely of the nature of foreseeing what is coming, but tend to be concerned with the consequences of possible courses of action.

GAMES, DATA BANKS, AND MODEL BUILDING

Underlying any major work in operational gaming or simulation for planning and control is the need for a data base and a conceptual scheme to guide the processing of the data. Large-scale data gathering may be necessary, requiring special studies. Cross-section or time-series data are often called for. Unless they are available from the operating information of some organization accessible to the investigators, the costs of data gathering become prohibitive in both of time and money.

There are a few simple benefits derived from the utilization of a computer simulation to provide the background for a game. These are so basic that they can be easily overlooked and yet are sufficiently valuable that they can often justify the work entailed in the construction of the simulation. The first is the construction of procedures for the checking of logical consistency and completeness in models. The second is the semiautomation of dimension-checking and consistency-checking among the basic data of the units of measurement. Once a system contains more than five or six variables, if no formal conceptual scheme exists that uses all of them, inaccuracies and inconsistencies can creep into data-gathering schemes. The use of a formal model guards against this. Simulation provides this degree of formality.

Both individual equations and the system as a whole must be checked. Before proceeding to the measurement of parameters and the estimation of equations, one must determine if the equation is at least consistent with the basic facts; the dimensions of both sides of an equation must match. Some fundamental errors can occur by not understanding the dimensions involved in the description of an equation. Some classical examples are provided by attempts to produce statistics comparing automobile deaths with deaths in air crashes. What is the measure? It is the number of miles per passenger, the number of trips per passenger, the ratio of successful flights to crashes, or some other standard?

50

Gaming, planning, data gathering, and model construction are allied but different occupations. There is a considerable gap in communications between practitioners. Methods of checking and validation which are routine to some disciplines are more or less unknown to others. The relatively simple and stringent rules which must be applied to formalizing a model for simulation can be of direct benefit in helping to bridge the gap between quantitative and qualitative approaches and between planning and research by providing a language more precise than ordinary language and yet far less restrictive than mathematics.

MEASUREMENT, VALIDITY OF MEASUREMENT, AND COST

A natural question to ask in the gathering of data during the building of any complex model is "How well can we describe and measure a variable or a parameter?" As we have already noted, the question in the context of gaming and simulation for planning or control is, "What are the costs and values of measuring this variable; and what are the gains associated with varying degrees of accuracy?"

A relatively complex simulation of the environment can be regarded as a preliminary research mechanism that, combined with the considerations derived from theory, may help to guide the selection of those aspects of the outside world on which the major expenditure of time and skill in measurement should be spent. This can be done by building the first models relatively quickly and crudely, using whatever statistics, special knowledge, insights, and theory that are easily obtainable. With this preliminary model, a sensitivity search on a few variables, parameters, or other aspects of the model deemed to be important may indicate where accurate measurement and further modeling are critical.

SOME IMPORTANT CONSIDERATIONS WHEN CONSTRUCTING SIMULATIONS

The uses and types of simulation may be different. In physics, a simulation may be used to calculate the behavior of many particles acting according to some random process. The overall behavior of the system is predicted by many replications of a Monte Carlo process.[1] It is possible to devise a Monte Carlo method to evaluate complicated mathematical integrals or approximate the value of the number π. These uses are called simulation by those who use them; however, they are far removed both in intent and technique from many of the major uses in planning or gaming, and in the behavioral sciences in general.

In the construction of simulations for large-scale gaming exercises or even for other purposes, there is a series of features that distinguish the different problems that can be encountered. A list of them is given below:

1. How easy is it to obtain a good representation of the system to be simulated?
2. Can an objective function, or some other clear criterion by which results can be judged, be easily or uniquely defined?
3. How important are random elements in the simulation?
4. Are logical switches necessary? For example, if the system is to include decisions with alternatives to be selected according to criteria depending upon indices whose values may change during the simulation, they will be needed. There might, in an international-development simulation, for instance, be two development policies, depending upon a level of confidence in political developments.
5. Is the number of variables and parameters so large that organization and machine problems will be critical?
6. Is it necessary to solve simultaneous relations within the simulation?
7. Is set searching and list processing of importance to the simulation? For example, in a simulation which may contain a set of firms (say, retailers) it may be desirable to locate all retailers in the Northwest who are above a certain size and who have more than k items of product 5 in stock. This involves the ability to classify special sets of firms and to search through larger sets, checking for fixed or variable attributes. There is a great variation in the types of computer language and programs to handle this type of activity.
8. How important is mass-data processing? There is a considerable difference between problems that involve relatively simple manipulation of large amounts of data (such as accounting schemes) and those that require the complex processing of smaller amounts.
9. Are relatively sophisticated mathematical techniques of importance? With maximization processes, or the need to solve simultaneous systems or differential equations, the mathe-

matical computational needs are relatively high: with many business-oriented models they are low.

10. Does the model need to be flexible? Are many changes, revisions, and modifications contemplated as a matter of course? For planning, experimental, and exploratory simulations in general, the answer would be yes.

11. Is the program intended for "on-line" use? In other words, is there an immediate feedback from the user to the program calling for immediate response? Ideally, in a short-term planning process using a simulation, the user should be able to interrogate the computer directly. There is a considerable difference in design between this type of program and a program constructed solely for research or operations.

12. How important is input-output and the appearance and flexibility of the format in which output appears? Do you need graphical, statistical, or tabular output?

The borders between science and art, the understanding of technique and of principles have always been ill-defined. Several of the points raised above appear to be purely technical considerations. For instance, most of the problems in number 12 in the above list sound as if they should be the concern of a typesetter and not a gamer or a behavioral scientist. Unfortunately, this is not the case. Considerations such as this are not sufficient to guarantee success, but they are almost always necessary.

THE ROLE OF COMPUTER LANGUAGE

The answers to the 12 questions asked above need to be considered not only in determining how a game should be computerized, but to help the designers select the computer language they intend to use.

The details of the description of different languages and the evaluation of their differences and efficiency for different purposes is of extreme importance as a means of cost control in constructing large games for simulation. This is a special technical subject of sufficient importance that if a game designer does not have the expertise himself, he must be able to use the services of someone who does. For those who wish to read further on this subject, there are excellent articles by Tocher,[2] Krasnow and Merikallio,[3] and by Kiviat;[4] a clear and simple exposition on languages in general is given in an introductory book by Barton.[5]

A computer language is a language that enables a user to talk with a computer. There are hundreds of computer languages; in some cases, they have been short-lived. The comparison of a computer language with ordinary language is most apt and useful. When humans communicate with each other, they do so by such means as writing, voice, pictures, signs, smells, touch, or gestures. Each one of these methods of communication might be regarded as a language. Computer languages to date have been constructed solely to be written, but in some instances there are picture inputs[6] which may also be accepted as part of computer language. In spite of the enormous progress made with computers, they are still far from able to replicate the subtlety and complexity of much of human communications.

Most human beings learn very few languages during their lifetime. To learn a language well requires a large investment of time. Computer languages are in general easier to learn than human languages; they are more precise and far more limited. In order to make them easier and quicker to learn, the mode of communication provided by a language such as FORTRAN IV between the human and the computer is more akin to a talk translated at the U.N. than it is to a conversation between two individuals.

The internal logic of a computing machine—its hardware design—determines the way in which information must be presented to it for operation. Users have neither the time nor the inclination to learn the extremely lengthy and complex combinations of symbols which go to make up the internal machine representation of the instructions they give the computer. This being the case, attempts are made to design programming languages that are especially easy to learn and are tailored to the needs of the user. These languages then go through one or more stages of translation before the machine actually uses them. There is, however, a cost as well as several gains incurred in this process. Specifically, the closer that a user language becomes to the user's ordinary way of expressing himself, the more translation this language will require before the machine can understand it well enough to operate on it. Albeit that the translators used to transform a user language to a machine language are programs themselves, they nevertheless cost money to run. Thus, in general, one trades user time for machine time and length of execution of a program. Furthermore, the less precise a user has to be, the more probable it is that some of his instructions may be misinterpreted.

A user-oriented gamer or simulator needs to distinguish three levels of language when dealing with computers. They are: special-purpose user languages, general-purpose user languages, and machine-oriented languages.

The distinctions among them are not always clear and there are invariably borderline cases. Other breakdown have also been suggested. Barton[7] suggests five levels of computer languages: (1) internal machine language, (2) external machine language, (3) machine-oriented symbolic language, (4) user-or problem-oriented language, and (5) user-written language. The last three are the equivalent to the three user-oriented languages noted above.

The internal machine language is the ultimate symbolic representation used by the hardware of the machine. In general, machines operate in a binary mode so that eventually any language is translated into a series of zeros and ones which can be handled and recognized electronically as "on" or "off" positions in an electrical circuit. Combinations of these two states enable a complete logic and mathematics to be built up. There is a trade-off in complexity between the machine language and the machine. Thus the computer specialist, if he does not like the way in which a machine handles a certain language, instead of modifying the language may modify the hardware of the brains of the machine.

It is possible to obtain copies of a program in a machine-operational form but these will be extremely lengthy series of zeros and ones. Their form is such that even a computer professional will find them inconvenient to read. For this reason, a simple first-order translation is made, in which machine instructions are represented by numbers or letters of the alphabet. In general, no ordinary user would be expected to read this external machine language.

The frontier between the machine culture and the user culture is reached at the level of machine-oriented symbolic languages. As the operations of a computer are painstakingly explicit, extremely simple commands that may need frequent execution may call for extremely lengthy and repetitive instructions at the machine level. For this reason, a level of translation is called for between the machine language and the user. By investigating the logical structure and the requirements, of both languages, it is possible to represent several machine instructions by a symbol or word in the symbolic language. This will be translated by a program (known as an assembly program) into full machine language. A machine-oriented symbolic language is closely related to the machine language that, in turn, is associated with a particular computer. Although a few English words appear among the string of letters, numbers and other symbols the uninitiated user will be able to glean very little information from a program presented in this language.

The user languages fall in a completely different category. They serve to cater to the needs and problems of the user first and they require further translation for the machine to be able to understand them. Some examples are COBOL, FORTRAN, and IPL. The first is oriented toward business problems, the second toward mathematical scientific problems, and the third is designed to help in the simulation of human problem-solving with a stress on list-processing. A program or translator, called a compiler, takes a user language and translates it into a machine language.

The two main features to be considered by a user concerning simulation languages are the modeling concepts and the implementation problems. A simulation language is a special language that concentrates on a particular view of the world. Kiviat[8] has suggested that simulation languages focus on four areas: transaction flows, events, activities, and processes.

GPSS has a flow-chart orientation and is natural and easy to use for engineers and others who are accustomed to building blocks of flow charts. SIMSCRIPT is an event-oriented oriented language. It is particularly useful in describing problems involving the scheduling of events. Although the article by Krasnow and Merikallio appeared in 1964[9] it presents an excellent example of how to compare languages. SIMSCRIPT, GPSS and DYNAMO are compared.

THE USER AND HIS PROGRAMMERS

Although a user can work quite successfully knowing virtually nothing about machine-oriented languages, it is of considerable importance to appreciate the nuances of user-oriented special languages. The problem facing even the most intelligent user is extremely complex and can best be labeled as a socioscientific problem involving the bureaucratic and technological, as well as the scientific, aspects of his work. For example, above a distinction was made between general-purpose and special-purpose computer languages. This distinction is by no means trivial nor is it always clear. From some points of view it might be argued that SIMSCRIPT is a more general language than FORTRAN, although one can find experts who would argue in either direction.

The individual who is problem: oriented and who does not intend to become a professional in the selection of computer languages is more or less a captive of his programmers and computational system, whether he likes it or not. His problem is to learn enough so that he can obtain an optimal tradeoff between the costs in terms of his time and the costs of his captivity.

Conditions in the state of computation have been evolving so rapidly that many problems of five or six years ago which were of critical importance are no longer so. For example, the size and the cost of program operation, the size of high-speed memory, and the speed at which external information could be obtained were all critical considerations in the construction of even moderate-to-small games and simulations some years ago. Currently, although economic calculations concerning size of programs and storage are needed, these types of limitations are important only for the largest of games and simulations.

Although the changes in technology, the growth in the numbers of computers, time-sharing networks, and the numbers of professionals all tend to solve many of the problems facing a user, he still needs to be aware of what the major problems are in the selection of languages. A few brief rules of thumb are suggested.

Know the problem and be able to express it relatively abstractly and concretely enough so that the programmers understand. This is closely related to understanding the "world view" or the basic modeling properties of the problem—whether it involves working with aggregate or disaggregate information, needs highly flexible input-output, will be modified frequently, or whether special events are important.

Beware of the ingenious programmer whose attitude toward programming is the same as an addict's attitude towards solving double acrostics. He may end up by saving a few seconds of machine time at the cost of producing an inadequately documented, incomprehensible program that can be operated only by him. Ask the question well in advance: what happens to the program after the chief programmer leaves?

Determine to what degree a special machine will be required in selecting a language being promoted by a computer center. Furthermore, it is also desirable to find out how long that center will maintain the system under contemplation.

Decide well in advance how generally the program will be used. For example, will they be run at a completely different center elsewhere? If so, a far inferior language that is in general use may be preferable to a highly specialized language that is available in the local center.

It is a relatively safe assumption that the best consultant is, in the long run, the most cost effective consultant. If you are involved in a large-scale gaming or simulation projects that is going to run into the hundreds of thousands or millions of dollars it is absolutely critical that the decision process to select a machine and a language be made explicit and be documented.

E X E R C I S E S

1. Describe the differences among special-purpose user-languages, general-purpose user-languages and machine-oriented languages.
2. For what type of simulation would you pick DYNAMO over SIMSCRIPT and vice-versa?
3. Consider the equation $X = AY$ which tells us that the number of traffic deaths is proportional to the number of automobiles. What can we say about the parameter A? Is it a pure number? If not, what are its dimensions?
4. Consider a weapon's evaluation simulation involving two types of dueling tanks and an international negotiation simulation. Contrast their different requirements in terms of their data requirements, model structure and the nature of their computer language requirements.

R E F E R E N C E S

1. McCracken, D.D., "The Monte Carlo Method," *Scientific American*, 5 (May 1958), 192.
2. Tocher, K.D., *The Art of Simulation*, Princeton, N.J.: Van Nostrand, 1963; Tocher, K.D., "Review of Simulation Languages," *Operations Research Quarterly*, XVI, (June 1965), 189–217.
3. Krasnow, H., and Merikallio, R.A., "The Past, Present and Future of General Simulation Languages," *Mgt. Sci.*, *11* (Nov. 1964), 236–267.
4. Kiviat, P.J., "Development of Discrete Digital Simulation Languages," *Simulation*, 8, 2 (1967), 65–70; and Kiviat, P.J., "Development of New Digital Simulation Languages," *Journal of Industrial Engineering*, 17, (Nov. 1966), 604; and Kiviat, P.J., "Digital Computer Simulation: Modeling Concepts," Santa Monica: The Rand Corporation, RM-5378-PR, August 1967.
5. Barton, R.E., *A Primer on Simulation and Gaming*, Englewood Cliffs, N.J.: Prentice Hall, 1970.
6. Groner, G.F., Berman, R.A., Clark, R., DeLand, E.C., "BIOMOD: A User's View of an Interactive Computer System for Biological Modeling (A Preliminary Report), Santa Monica, Calif.: The Rand Corporation, RM-6327-NIH, August 1970.
7. Barton, R.E., *op. cit.*
8. Kiviat, P.J., Development of New Digital Simulation Languages," *op. cit.*
9. Krasnow, R., and Merikallio, R.A., *op. cit.*

4 Costs and Procedures

Gaming can either be extremely expensive or it can cost virtually nothing. Costs are an extremely important factor in making decisions on the use of gaming. In keeping with the taxonomy of different uses of gaming noted in Chapter 2, costs will be discussed under several different categories emphasizing, for the most part, gaming for operational purposes, experimental purposes, and teaching and training purposes. Thus we are interested in the costing of games which are played in academic settings, governmental-diplomatic settings, military settings, and games which may be run by citizens, groups, or other organizations. Most groups utilizing gaming have tended to be relatively cavalier in estimating the cost involved.

In particular, in an academic setting the relationship between cost and scholarship is often played down or ignored. Even now there lingers in academia a feeling that costs do not constitute a proper question with which to confront scholars. Recently, however, with the financial crises in both teaching and research the vital question of costs has been raised.

With the growth of the technology of television and computers, new methods for teaching have become available. They must be contrasted with the old. Individual tutoring, for example, may be highly effective as a method of undergraduate instruction; yet, when the size of the undergraduate population is considered, it becomes an overwhelmingly expensive method of education. What are the costs of education per hour in terms of time of the instructors and students and in terms of the physical resources required? If the old methods of education are beginning to encounter serious economic limitations, what are viable alternatives?

Similar cost problems exist in areas such as mental health. Even if the current methods of psychoanalysis and psychotherapy were utterly successful, is such an approach economically feasible for a society with hundreds of thousands of patients, each requiring several hundreds of man hours of expensive professional time for this type of method to work?

In the costing and the measurement of the work involved in gaming exercises, we must distinguish between mass-produced games to be played by large numbers of individuals and very restricted games to be played by a few chosen individuals. The problems encountered in the costing of experimental, operational, or other forms of gaming are not dissimilar to the general problems of costing that are faced elsewhere. Mass markets cannot effectively be serviced by high-priced hand-crafted methods. Specialized needs for games with high payoffs but extremely limited use can easily call for painstaking hand-tailoring of games. In considering the cost-effectiveness of games, great care must be taken in examining the importance of the different needs they may attempt to fill.

Even the most insensitive of decisionmakers or behavioral scientists is soon overwhelmed by the complexities encountered in studying human behavior. In the simplest of experimental games, there are thousands of interesting hypotheses or conjectures which might be tested; in operational exercises there is a proliferation of possible changes in the environment or in policies which appear to be interesting, especially to those not running the experiment or using the game. The need for experimentation on human decision-making processes is pressing; the number of worthwhile topics for investigation is infinite; and even under the best of circumstances the costs are immense. The demands for teaching and operational purposes are neither smaller nor less pressing than those for experimentation.

The reports of individuals concerned with gaming indicate expenditures ranging from a few hundred dollars (with much time and effort not accounted for) up to many millions of dollars to construct and run elaborate operational games to provide insight on a relatively narrow range of military problems. The search for resources is exemplified in the use of students, prison inmates, or enlisted men as game players. Occasionally, ambassadors and generals may be used in a political-military exercise, in which case the play cost of the run may soon become staggering.

COSTING AND COST EFFECTIVENESS

A useful breakdown for cost-analysis categories in most gaming activities is as follows:

Research and Development Costs
1. game design and preconstruction testing
2. game construction and programming
3. debugging, test running and evaluation
4. major game modification
5. laboratory design and design of laboratory modifications

Investment Costs
1. laboratory facilities, buildings
2. equipment
   a. computers (including peripheral)
   b. other special instruments (audio, visual, etc.)
   c. office equipment, furniture, etc.
3. stocks
   a. Library
   b. Supplies
   c. Spare parts
4. personnel training
5. Miscellaneous
   a. travel and transportation
   b. other

Operating Costs
1. maintenance and replacement:
   a. laboratory and buildings
   b. specialized equipment, computers, etc.
   c. game specific equipment
2. salaries and payments
   a. direct salaries and wages
   b. payments to players
3. services and miscellaneous
   a. transportation and travel
   b. other variable costs
   c. other overheads

Research and Development Costs

Included in game design and preconstruction testing is also the design of the analysis program and data bank that may be required for the game. Implicit in the category of game construction and programming are also the construction and programming of the analysis system and the structure of the data bank. In many games these features are not relevant. Furthermore, in many games, formal laboratory facilities are not used. Nevertheless, some sort of facilities are used and it is important at least to know whether they are being obtained at zero cost.

The research and development costs may be regarded as those expenditures aside from equipment that are needed in order to bring a game to the stage where it can actually be played for the purpose for which it is being constructed.

Investment Costs

Investment costs are those necessary in order to implement the use of the game. The actual expenditures will usually be made after most of the research and development has been done, in order to avoid expensive modifications while in construction due to changes in research and development design.

For many games laboratory facilities and buildings are out of the question. Borrowed or rented facilities and equipment are frequently used, and the computer is sometimes rented. Often several games may be run at the same laboratory facility. This brings in the problem of the assignment of joint costs, which will be discussed later. A large-scale gaming operation may require a library and special forms and supplies. If the gaming operation is maintaining its own laboratory it will also require spare parts.

An important item often overlooked is the cost of training personnel. In academic settings, when graduate students are serving out an apprenticeship this cost could conceivably be zero or even negative (in the sense that the time they spend working can be allocated to the teaching budget). However, in private industry or in government the training of personnel for a gaming team involve real and frequently high costs. In large bureaucracies these costs may be hidden because of the "pirating of hours" from other projects by the group building the game.

In the opening up of a gaming facility there may be costs associated with the transportation of equipment and personnel. In general, however, these costs tend to be quite small. They are listed, however, to cover the construction and investments in installations such as the gaming laboratory at the Systems Development Corporation or the Naval Electronic Warfare Simulator.

Operating Costs

Operating costs include maintenance and replacement costs, salaries and payments, and services. With specialized equipment, especially computational equipment, systems can become obsolete at a great rate. Thus, even though no further developmental work is being done on a game, occasionally an equipment change is forced by the inability to get service or spare parts, or it is forced by other joint users. In many instances there may be game-specific equipment that is quite costly. For example, some games may require elaborate analog equipment built specially for the purposes of a single game.

Salaries and payments must be broken down into (1) direct salaries and wages received by those who run the game or the gaming laboratory and (2) the payments made to players. An easy source for the underestimation of the costs of a gaming exercise comes in ignoring the value of the players' time if it happens not to be directly charged to the gaming operation. For example, in a political military exercise, several ambassadors may spend two or three days playing a particular game; their expenditure of time may not be included in the costing of the game.

Under services and miscellaneous, transportation and travel may be extremely high if a game is to be played at more than one location, which involves transporting a complete crew and equipment. Other variable costs include office supplies and incidentals. A category called "other overheads" encompasses unforeseen expenses.

PROBLEMS IN COSTING

Often it is difficult to predict how much money to allocate to a particular category due to the environment in which many games are run. Even a quick calculation will show that the simplest of controlled experiments examining only one variable cannot be performed without using up several hundreds of dollars worth of resources in time and/or equipment. The casual and informal attitude of teachers, experimenters, and students toward their use of time may disguise many of the costs. It may take three or four days to set up the experiment or make instructions explicit; even if this time is not reported, the costs have still been incurred. Even though volunteer players might have given their time freely, the report that a certain game cost $150.00 to run does not mean that it can be replicated at anywhere near that figure in another institution or laboratory.

Games have been run in which the imputed costs have ranged anywhere from next to nothing up to several million dollars. As has already been indicated, the variety of types of games and their purposes is considerable. For many games, the major cost is time—that of both the players and the administrators. For others, the most important cost consideration is the computer. Another major cost factor concerns facilities (this can include the computer costs). Information on facilities can be relatively unsatisfactory as the charges are often based on accounting fictions which arbitrarily allocate joint costs or charge members of the same organization for the utilization of resources whose alternative employment is zero; this applies to the cost of computer time as well. Long-run costs are very different from short-run costs, and so far not enough attention has been paid to examining the length of usage of a game.

Frequently there are alternatives to gaming for experimental teaching, operational, or other purposes. The choice to employ gaming must depend on its effectiveness and its costs relative to the other methods which could be used. Anyone with a serious interest in the methodology or the theory of gaming must consider its cost aspects and their implications. Possibly the two most important features which must be taken into account by a game designer concerned with cost are the time distribution of cost and the vagaries of joint allocations or unimputed allocations of cost. In the next two sections both of these features are examined.

The Timing of Game Construction and Use

In the discussion of costs noted above, costs associated with the design and construction and maintenance of a laboratory have been categorized. The laboratory and equipment costs of gaming form a separate but important subject. In this section observations will be limited to game construction and use.

Assuming that it is possible to attach a monetary cost to all of the aspects of the resources invested in the construction and use of a game, a highly useful planning device is a histogram differentiating operations costs, investment costs, and research and development R and D costs. Figure 4.1 shows a hypothetical case. When costing games, much of the investment cost can be the overhead assigned to pay for a share of an existing laboratory. However, in some instances the investment cost may represent a direct cost of special equipment and materials for the game at hand.

The histogram drawn in Figure 4.1 could represent the costs incurred in the careful use of an operational game, such as a politi-

64

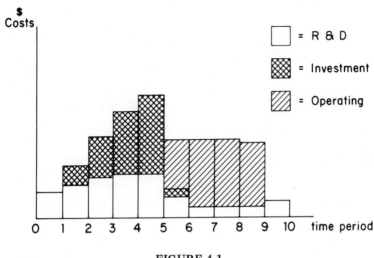

$
Costs

☐ = R & D

▨ = Investment

▨ = Operating

0 1 2 3 4 5 6 7 8 9 10   time period

FIGURE 4.1

cal-military exercise. The first buildup costs are primarily R and D. On top of that comes investment in special equipment for the particular game. For example, special data banks might have to be constructed, and a certain amount of peripheral equipment, including items such as cameras and tape recorders, might be charged to the specific game. By the time the game is ready for operation, most equipment will probably be around. However, there may be a period of "debugging" where last-minute arrangements are still being made and there is interaction among operating features of the game, investment, and research and development. In Figure 4.1 serious playing begins between the fifth and sixth periods. Investment costs have dropped to a very small amount, while R and D costs, although smaller than operating costs, are still moderately high. The next three periods represent the intensive use of the game, where operating costs represent almost everything. However, in a well-managed game there will still be some R and D costs resulting from interpreting the results of the game. After the ninth period the game is no longer in use; however, it is important to observe that R and D costs are marked as having grown for this period. A sensibly used operational game will be followed by evaluating, recording, and interpreting what was learned from the game, not only from the viewpoint of the game itself, but from the viewpoint of improvement of future techniques. In practice, frequently the last step is not taken.

Figure 4.2 presents a flow diagram of the design and use of a game broken down into: R and D, data and setup work, running of

65

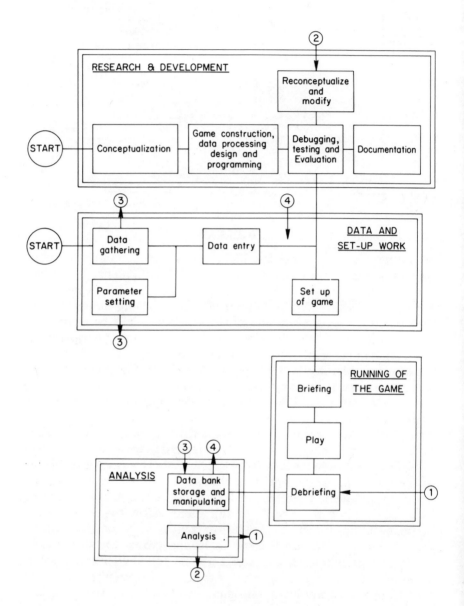

The symbol → ① implies a connection to ① →.

FIGURE 4.2

the game, and analysis. Frequently documentation is not attempted until after the game has been in use for some time; however, it is a safe and sound procedure to start documenting early in the game and update the documentation along with the operational use of the game.

A partially sequential, partially parallel process to the construction of a game is data gathering. Until a game has been somewhat formulated, it probably does not pay to do too much data gathering. However, at some period, according to the specific type of game being constructed, it will usually pay to start the data gathering parallel to the game construction. In operational games, where specific scenarios relating to actual world conditions are being used, the data-bank and data-gathering problems are of considerable size and much time can be lost if parallel work is not performed.

Large, complex experimental games, operational games, or games for teaching and training may require elaborate parameter settings. In cases such as this, it is desirable to have work done on preparing parameter packages so that games can be set up quickly and if necessary might be repeated, or there is at least documentation for comparative analysis.

A process that is frequently overlooked is that of data-handling and analysis. Part of the design of any new game should be the design of the analysis and data-handling of the output. An efficient gaming center will be concerned with the analysis of the individual game and with its relationship to other games that have already been run. As part of the laboratory equipment there might be many programs that are used for more than one game.

A safe rule to follow when considering analysis packages for experimental or other types of gaming is "Beware of methodologists bearing general purpose programs."

After the analysis for a game has been performed, it will serve as an immediate input to both debriefing and to the reconceptualization and modification of the game. In the actual running of a game, there are three divisions that are usually worth making. They are the briefing, the play, and the debriefing. Games appear to be far more sensitive to briefing than most of the literature acknowledges. Professor John Kennedy has noted in personal communication half-seriously, "Tell me the results you want and give me control of the briefing and I will get you your results." Depending upon the use of the game, it is important to link the debriefing with the analysis. For example, at the end of an elaborate operational game, there might be a lengthy debriefing, usually done on the last day of play or the following day. However, it might take several days to interpret what actually happened during the game. Furthermore, it might be advisable to perform detailed analysis of some of the output. For the most part, this type of analysis or postmorteming of the game is rarely done. Furthermore, if it is done, it is rarely given back to the players, as they tend not to reassemble after the immediate debriefing.

These comments on Figure 4.2 are designed to present a quick overall view of the component parts in the design and use of a game. Emphasis has been given to timing. In the next section, some general observations concerning the difficulty of costing are made. Subsequently, the different components are examined in somewhat more detail from the viewpoint of resources, costing, and timing.

Allocation of Cost and Opportunity Cost

There are four major difficulties in the costing of a game. They are: unallocated costs, opportunity-cost evaluation of special resources, overhead and joint cost allocation, and the confusion of costs with benefits.

The group most likely to underestimate resource costs by failing to include charges for them are academic gamers. They are most likely to fail to assign costs to R and D time, player time, and computer time.

Opportunity costing is an important factor in avoiding distortion of the costing of a game. It is particularly important when one is dealing with resources that are special and resources which cannot be moved. In particular, the concept of opportunity costing is critical for the adoption of an economic viewpoint of an activity, yet at the same time it is not appreciated or is frequently misunderstood by even extremely highly trained technical personnel.

A brief anecdote can perhaps help to explain the key concept behind opportunity costing:

Johnny goes to school and participates in a spelling competition. He wins the competition, and the teacher offers him either an apple or an orange as his prize. He selects the apple and goes home proudly carrying the apple. His mother upon seeing him carry the apple says to him, "Johnny, how did you get that apple?" Johnny replies, "By not taking an orange."

The opportunity cost of the apple was an orange. The opportunity cost of most items in well-defined markets is the amount of money one can obtain for them which can then be used to purchase anything at that price. However, frequently items are indivisible or are not up for immediate sale. An item such as a deserted classroom in an evening is usually not necessarily either for sale or for rent. The opportunity cost of a classroom can frequently be accurately assigned a value of zero. If the gamer does not use it, then it will remain empty.

It is important to distinguish the difference between assigning a zero opportunity cost to a physical input and ignoring the costing of the input. In the first case, because the resource was explicitly enumerated, if the game were replicated elsewhere, the need for the resource would be explicit and it could be easily determined whether that resource were free in the new circumstance. If recording the zero opportunity cost assignment were ignored, the costs associated with the game could be seriously underestimated if it were to be utilized away from its present location.

In costing personnel used in a game, several delicate organizational problems are encountered. The easier and two more common methods of costing are either to attach no cost to the personnel, or to count the numbers of hours involved and to directly charge that proportion of each individual's salary. Neither of these methods is completely satisfactory. In many instances, at a university, in the armed forces, or in a research institute, individuals are permanent staff members. This means that the assignment of wages may not necessarily be a good reflection of the worth of the individual time invested in the game, as a few hours might be regarded as overhead. Empty classrooms, borrowed laboratory facilities, second- or third-shift computer time, and the time of personnel are all items which under certain circumstances should be looked at in terms of opportunity cost.

In large bureaucracies there are invariably large overheads, and the assignment of overhead costs usually has a large degree of leeway attached to it. This is especially true when an operation needs statistics from ongoing parts of an organization. Depending upon the whims of the administrators, these inputs may either be supplied at no charge or the requesting group will discover that they are paying for virtually all of the data processing done by the components supplying the information. One cannot totally disassociate the problem of opportunity cost from joint costs. The connection comes up in situations such as who should pay what share of a jointly used laboratory facility.

The protagonist or antagonists for the use of games frequently confuse costing with the benefits or the product of a game. For example, the author believes that experimental games when attached to educational or entertainment uses can actually be run at a positive net income. This means simply that revenues will exceed costs. When proponents of gaming argue against costing the participation time of top level personnel because they gain more than the time is worth, this may be true, but it amounts to a confusion between costing and the evaluation of product.

The Purposes of Costing

It is important to remember that an accounting system or a costing method is valid only with respect to a set of questions it is meant to answer. It is an input to a decision system and should be treated as such. Costing to minimize income tax is a different procedure from costing to conform to bureaucratic requirements of a grant. Costing to attack a project as being a waste of money because of inefficiency is a different process from costing to "whitewash" a project that might have wasted large resources. Costing to calculate whether or not it is feasible to perform a piece of work within a specified project is a different process from costing to justify a work program.

In particular, the individuals interested in the use of games for teaching, training operations, or experimentation should be extremely conscious of at least three forms of costing. The first is merely a listing of the physical resources used. This has already been discussed in the three major categories given for the breaking down of inputs. The second is the assignment of monetary costs to those inputs to reflect the cost of investment, R and D, and operation of a game at a specific location over a specific period of time. The third is determining the cost of running the game elsewhere. This calculation will be dependent upon the facilities and the mobility of the facilities available in the new location.

GAME CONSTRUCTION

Gaming in most of its uses, in spite of a venerable pedigree for some war games, is a new activity. The number of professionals in gaming is relatively small. Except for the larger military games, some political military exercises, a handful of experimental games, and a few of the games constructed for teaching and training purposes, most games have been constructed in a highly informal manner, usually with a few interested members of a research organ-

70

ization meeting informally and pirating resources from elsewhere.

Game construction offers one of the most important teaching, training, and operational uses of a game. Game construction is an exercise in model building. One of the most important lessons that constructing a game has to teach is in abstraction and conceptualization. If one is studying urban redevelopment or foreign policy in a particular area of the world, the discipline involved in constructing a reasonable, playable game is probably worth more than the actual playing of the game. For this reason it is desirable to use many participants or assistants in the construction of virtually any game.

The above observations are especially true for operational games and for games used for teaching purposes, especially at university levels. Many of the current courses in operations research, management science, political science, urban development and other behavioral sciences call for the construction of the abstract models of organizations and decision-making processes. The construction of a game provides a rigorous way in which a student can be brought to a clear formulation of his ideas. It further presents a format in which he can search his model in a direct manner, both to locate logical flaws or gaps in its construction and to obtain insight into its realism and relevance.

Game construction is closely related to the general problem of model building and to systems analysis. For a game of merit, four skills are required in construction. They are:

1. the ability to formulate the appropriate abstract representation of the phenomena to be considered
2. the skill in the construction of viable model, i.e., one that is amenable to control, to manipulation, and to analysis
3. programming skill, especially with respect to modifications of the main program and flexibility of input and output
4. the ability to design data-processing procedures for output and, in some cases, design experiment

The first skill requires a good substantive knowledge of the type of phenomena to be investigated. The discipline involved in the construction of a well-defined and relevant model is great. The second skill calls primarily upon methodology and technique. In the course of time, individuals learn what are efficient approximations, manageable functional forms, shortcuts, and special procedures. Unfortunately, at this time, there are no handbooks and few books or articles which contain any detailed techniques of game construction.

Programming skill, a few years ago, was a sufficiently scarce commodity that its shortage presented a bottleneck. Currently the growth in the availability of programmers together with the improvements in computer languages has removed the bottleneck. Possibly the most underdeveloped and difficult area in gaming of all types is in data processing.

For many purposes the cost of game construction can be completely avoided by using someone else's game. Certainly for teaching and training there are considerable incentives in using a few well-constructed games rather than having everyone build his own. The case is not so clear for operational games, where frequently a special situation is being studied. For teaching at levels below the university it is probably highly desirable that the resources available are spent on the construction of a few well-planned and frequently modified games rather than on many different games, each built with extremely limited resources.

One factor which seriously limits the feasibility of using games designed by others is the degree of documentation. The best documentation is usually associated with commercial products, such as games for entertainment. Even there, however, the documentation is usually only designed for instruction in how to play; for other games documentation is also required to indicate how the game should be analyzed and evaluated after it has been played. Furthermore, when games use computers, a great deal of technical documentation is called for. This is even more important when one wishes to encourage other users to modify the programs they have been sent. In Chapter 5 some information is provided on the development time and cost of a few laboratories and games.

In summary, a simple game for university-class use might take a few man-days of modeling; a moderate business game might take from a few man-months to a man-year of modeling and programming and from three months to a year before it becomes operational; a large business game, an urban development or international relations game, or a military operational game might require from two or three to several dozen man-years of modeling and programming and from nine months to two years before it can be used.

Experimental games may take anywhere from a few weeks to several months to design and from a few weeks to several months to program. Teaching games for preschool, public school, and high-school levels can and have been put together both cheaply and quickly; however, there is little evidence on the relationships between construction time and the value of the game. In other words, it is not hard to design quickly a game which may amuse

72

a certain number of children for some hours while they are at school. The problem that remains is whether anything of worth in terms of the goals of education could be achieved by the use of the game.

DEBUGGING

During the last 10 years there has been a revolution in the techniques for locating errors in computer programs. The improvement in computers, special equipment, computer languages, and special diagnostic routines, together with the growth in numbers of trained computer personnel, has been so rapid that locating programming errors and meeting the specifications of different machines, problems of considerable economic importance a few years ago, can more or less be ignored now.

In big operational games, where the model may represent a large logistics, command and control, or defense system, the program may become so large and intricate that validating the design logic of the basic model and the program logic becomes a task of immense proportions. Even in a large business game, such as that used at Carnegie Mellon, or the Fame game constructed at IBM by Levitan and Shubik, "bugs" will be discovered several months, and possibly years, after the game has become operational. The existence of "bugs" is closely related to the quality of documentation and size of the game. In general, the quality of documentation is poor for games produced by academics, and somewhat better but still surprisingly poor for research institutes, the military, and commercial enterprises. In most games used for graduate teaching purposes, even if they are complex, debugging does not represent a serious problem. Much of the value in their use (together with a certain amount of the joy) comes when the users find errors in the logic or flaws in the modeling.

In "free-form" operational games, such as those used to structure the investigation of a political or diplomatic situation, or to check the effect of new tax laws, the rules are not completely defined to start with, and much of the use of the game is in the exploration of what has been missed or misanalyzed in the original formulation. Most of the experimental games are sufficiently simple in their construction so that a hundred percent validation of the design logic and the program is as feasible as it is desirable.

In general, it is safe to assume that if an error or omission is discovered in a game designed for computer operation, it is in the long run cheaper and easier to reprogram the game or part of the game so that it operates in exactly the way desired than it is to patch up the program with odd-hand corrections and such features as extra, out of order, nonnumbered input cards, or handwritten additions to output formats.

The importance of "good housekeeping" is frequently overlooked, especially by those who construct games somewhat informally. Cutting corners on trivial items, such as office supplies, keeping corrections on scraps of paper, failing to correct small errors because one intends to remember how to correct them by hand, in general, result in eventual disaster for virtually any gaming use.

TRIAL RUNS AND DRESS REHEARSALS

For most teaching purposes trial runs are both necessary and relatively cheap and easy to run. The major costs are usually the time of personnel. If designed well, the trial run will not only help debug the game but will also help to train personnel for the control team or group using the game.

In some free-form operational games, dress rehearsals pose a problem as they may require the use of relatively high-level participants for periods from several days to several weeks to "shake down" the game, check out the briefing, and run through the "choreography" before the game can be played, without running the risk of incurring the hostility of the participants with an unseasoned game. In large operational games, trial runs or dress rehearsals are both necessary and relatively expensive in terms of manpower and computer time. Experimental games require trial runs, and, for the most part, give rise to problems that are not faced in the other uses of gaming. Especially when one is dealing with an environment-poor or "dust free" situation, the results of a game may become very sensitive to the type of briefing given to the players.

For small, informal gaming exercises, administration causes few, if any, problems in any uses. For large games, the degree of computerization is an important factor. Large noncomputer games, such as the Internation Simulation at Northwestern University, run by H. Guetzkow, in its original form required many assistants, an efficient filing system, and people and space for hand calculations. For any large game, computerized or otherwise, a laboratory manager becomes a virtual necessity. With large operational games and with mass experimentation, the logistics problem of catering to the needs of a large number of transient personnel can quickly become time consuming and expensive.

Especially critical for experimental and operational gaming is the transcribing, organizing, and storage of output; modifications to programs or experiments; special subroutines, and so forth. There is a great danger in building up catacombs filled with unanalyzed and uncoordinated information. There is also the danger of its becoming an institution so quickly that control over it is lost. This happens when ingenious, but poorly documented corrections are made or subroutines are added, and when many individuals are working on the same game without an appropriate amount of coordination.

DOCUMENTATION AND REPLICATION

If games are to be used by more individuals than just the originators, it is critical that they be well-documented. Good documentation is expensive both in time and money. For a game to be played at several locations without the previous users or builders present, good documentation is necessary. Furthermore, if a game has been made into a broadly distributed commercial product, it usually is well-documented.

In evaluating the degree and use of the documentation of a game, consideration should be given to the existence and availability of:

1. the program listing
2. flow charts, variable listings, and comments on the program
3. the program deck or tape and special library routines
4. a computer operator's and programmer's manual
5. a game manager's manual
6. a player's manual
7. analysis routines and debriefing instructions

The compatibility of programs written for one computer with the system of another is also important.

If a game is not to be used in more than one location, relatively informal documentation may suffice. If it is to be used in more than one location, the added documentation necessary may easily present a burden, because it is in fact a joint cost and may often be borne by only one institution. A further important deterrent to documentation is the demand on the time of senior gaming personnel, the designers of the game, and the experimenters. Funds may be available for secretarial and editorial help, but the important bottleneck is the availability of time of the top professionals.

One indication of the maturity of a science is the ease with which an experiment can be replicated elsewhere and the same results obtained. This author has been unable to discover more than a very few instances in which a gaming experiment has been replicated and like results obtained by others than the original experimenter. One of the reasons for this is the difficulty in controlling the effect of briefing, even under relatively good experimental conditions; another is that documentation is generally not explicit enough for the experiment to be a replication without the author standing by.

Sometimes a game being run at one institution appears to fulfill the needs of another institution. A safe rule to apply before using someone else's game is that if the original game requires extensive modification and is poorly documented, it soon becomes more economical to build a new one. There is an extremely difficult problem in administration and cooperation that has to be faced by game users. Costing is difficult enough in a single institution, but when there is more than one institution involved, the sharing of joint costs and the coordination of work become extremely difficult.

The production of games for entertainment encompasses several relatively well-organized industries—producing puzzles, board games, and sports events. In particular, those interested in the use of games for teaching, operations, or research have much to learn from the managerial control and the organization of games produced for commercial purposes in the areas of production, distribution, testing, and documentation. In running contests and in training, a lesson can be obtained from observing organized sport.

THE COMPUTER

Gaming for most purposes is becoming more and more computerized. There are several important reasons for this development, the most important of which have been the decrease in the costs and increase in the availability of computers, coupled with the growth of gaming technology and the number of individuals acquainted with the new technology. With the advent of time-sharing, it becomes somewhat more difficult to discuss the computer requirements without immediately discussing laboratory facilities. Laboratory facilities are detailed in Chapter 5.

There are eight factors in considering computerized vs. noncomputerized games:

1. costs and availability of computers and trained personnel
2. importance of computational error in the game
3. administration and paper-handling costs
4. importance of controlling time delays and permitting different levels of interaction
5. displays and input-output instrumentation
6. need to automate analysis
7. importance of data banks and parameter control
8. mobility

It is likely that the costs of computation will show the same spectacular reduction in the next five years as they have in the last; furthermore, there are other long-run economies which are beginning to be manifested. Computer languages are still improving considerably, and the chronic shortage of competent programmers has eased.

In many operational and experimental games the accuracy of computation becomes an important feature. Whenever the game structure is anything other than extremely simple, error and/or delay becomes virtually certain in hand-calculated games. The consideration of error generation alone is sufficiently important to make computers a necessity.

The administration and paper work in the running of many games involve not only considerable organization but considerable manpower. When research assistants are cheap or free, space is ample for them not to get in each other's way, and turnover is low enough to ensure the continued availability of a game administration team after it has been trained, then a noncomputerized labor-intensive operation is economical. If these conditions are absent, the automation of the paper-handling will represent an important saving.

The sensitivity of a gaming exercise to time delays is another factor. These delays are often caused by the correction of hand-computation errors or delays in the physical process of having research assistants act as messengers.

Balanced against the causes of time delays in a noncomputer game are the possible difficulties in running a computerized game utilizing the services of a computer without control over the priority system. Fortunately with the growth of timesharing the difficulties in scheduling have been considerably reduced. Unfortunately, the problems of ownership between a large center such as a university computing center and an individual experimental gamer are still great. A large user may feel that he is being asked to bear an unreasonable amount of overhead costs. Furthermore, he may find that he cannot get the configuration he wants. In some instances, this may lead the individual to abandon the joint, centralized computer center and make arrangements to obtain a dedicated machine for his own purposes. This is an economic decision which depends upon items such as the way in which joint costs are being allocated and the problems of committee management by a multitude of users with different orientations.

Display devices and input-output methods and instruments are features of computer systems that are seeing considerable growth. New programming languages are facilitating the generation of input and output formats. Improvement and increases in the variety of peripheral equipment have made it possible to completely automate games to the extent that not only may several players enter their decisions directly into a console and receive messages back in a variety of displays almost immediately, but further-automation of data processing and of the general housekeeping involved in recording experimental information in data banks, and the use of computers in an interrogative mode are all within the realm of possibility. The use of a computer in an interrogative mode means that the user can question it whenever he encounters a difficulty and has a reasonable chance of obtaining an immediate answer.

For simple experimental games, as well as in teaching, time-sharing by the direct inputting decisions and/or other information permits a considerable increase in speed and efficiency of operation. It allows the subjects to attain their own desired fastest speeds of operation and thereby eliminates many of the important difficulties associated with boredom. Automated small classrooms and games have been successfully run at the Systems Development Corporation, the IBM research labs, Berkeley, U.C.L.A., Yale, and elsewhere. However, there are still many unsolved administrative and scientific problems involved in time-sharing on large computers. In spite of the difficulties with time-sharing, however, the improvements in other aspects of input and output, such as increasingly flexible formats and the availability of better graphical displays, make gaming administration far easier and more effective.

An open question concerning the computerization of inputs and outputs is its effect on the game. In some games the players are allowed to design their own information flows. Although it may be cheaper to have a player type decisions directly on a console and also obtain a typed out message from the machine, not enough is known about how much changes in input and output devices affect behavior.

Possibly the most important use of computers to science in general and gaming in particular is in data reduction. Statistical tests which were inordinately time-consuming without computer assistance can now be run routinely at little cost. Sometimes it is possible to modify existing programs so that the costs of developing analysis packages can be kept low. Most of the games currently utilized for teaching have not had their outputs subjected to analysis further than that performed in the debriefing sessions. It appears that many games used for teaching for operational purposes or even for entertainment could, with slight modifications, produce experimental data.

The use of a computer for analysis is almost, but not completely, distinct from using a computerized game. If the game being employed is computerized, then routine data organization, generation of tables and graphs, and the coordination of raw information concerning inputs, outputs, and running conditions are supplied almost without cost.

Large-scale gaming for teaching or for experimentation is relatively new. There is still a dearth of trained and interested individuals. Hence, to some extent, at this time, the observation that almost all games being used for teaching as well as research could and should have an analysis program to accompany them is merely speculation. Nevertheless, we must evaluate the possibility of a considerable savings in cost by obtaining experimental data as a biproduct of a teaching activity (and possibly vice-versa).

The role of the environment in a game is heavily dependent upon whether it is a free-form or a rigid game. The more free-form the game tends to be, the less amenable it is to computerization. In some instances realism may be deemed to be an important factor— even more than, say, control.

In diplomatic exercises, the referees may pass on the realism of moves. In rigid gaming where the rules are all well-defined from the beginning, a computerized game provides for both a considerable flexibility and for a degree of control that is not easily achieved with a noncomputerized game. An appropriate set of parameters and the availability of alternative subroutines make the games easy to modify. Furthermore, in some too-rare instances, sensitivity analyses and exploration of the response surface can be made inexpensively, thereby giving the experimenter additional control over the game.

Several attempts have been made to construct dummy players. If the game is rigid and relatively simple, it may be possible for the experimenter to writer down a strategy for the dummy. Once the game becomes relatively complex, it is necessary to write a program to simulate a dummy player.

The last point considered in comparing computerized to noncomputerized games is flexibility in playing arrangements. Simple noncomputerized games, like Monopoly, Management, or Summit, can be played almost anywhere with very little equipment. Complex noncomputerized games, such as Guetzkow's Internation Simulation, need a great deal of office space and may even require a formally laid out laboratory. Large computerized games may need both laboratory and easy access to a computer.

A Warning on Time-Sharing

Time-sharing encompasses many different phenomena. The type of time-sharing system needed for experimental games differs somewhat from that needed for operational games. Furthermore, they differ totally from the time-sharing system that might be needed by a librarian or by a physicist.

It is safe to advise skepticism toward the individual who promises a time-sharing system able to solve all of the problems of a diverse group of users. The only guarantee of reasonable performance is to know one's own problems in detail and then to find a computer expert who does not have a particular system to sell. General-purpose systems tend to be no-purpose systems. This is frequently discovered after one has committed all of one's resources.

BRIEFING AND DEBRIEFING

Briefing refers to the preliminary session before playing the game, in which the players are told the purpose of the game and are provided with instructions for playing it. Questions are answered at this point. For some experimental games preliminary tests may be administered to the players during this period to check their understanding and perception.

The quality of the briefing is most important to operational games and to experimental games. It is somewhat less important for teaching games and entertainment games. There is more opportunity to correct the effect of poor briefing during a teaching exercise than there is in an experimental or operational game. Beyond understanding the rules of a game being played for entertainment, the individual need not gain too much from a briefing. In fact, generally no formal briefing is given; the player reads the rules or already knows them before playing the game.

Debriefing is the session that follows the play of the game, in the case of an operational game, the players and the control team evaluate the lessons learned from the game. In experimental gaming, debriefing may not be as important, but in some instances it may be desirable to administer a battery of tests to the players after the game is over. Debriefing for teaching games is usually not performed, although it would appear to be useful to at least use some tests to measure the effect of the game on the players. In a formal sense, a debriefing is rarely if ever held in association with games played for entertainment. However, it is worth noting that often individual players enjoy the "postmortem" more than they enjoy the game. This is especially true in games such as chess and in board games simulating specific battles.

The design of careful briefing and debriefing methods is a time-consuming and highly professional occupation. For operational games it is tied in closely with the design of the scenario. For experimental games the briefing is frequently the key to the results to be expected.

Debriefing cannot be reasonably disassociated from the analysis of the game results. A disturbing feature of the use of operational games to date has, however, been the surprising lack of linkage between debriefing and the analysis of the results of the game played. Frequently when the game is played there is a debriefing session for about half a day. The results are then shelved or thrown away. There is rarely a follow-up debriefing making use of a serious further analysis of the play.

ANALYSIS AND DATA HANDLING

A specific game may be looked upon as a data-organizing device. The questions to be answered are an important guide to the data to be gathered. A well-designed game frequently supplies a well formulated set of questions, and hence serves as an organizing device for the data bank. In other words, an abstract model or a simulation serves as a basis for well-defined questions. Hence, it

81

provides a natural way for organizing a data bank. Logical consistency and a reasonable amount of completeness combined with the right level of aggregation and an understanding of what is a relevant and well-formed question are keys to successful data gathering and manipulation.

EXERCISES

1. Describe a breakdown of gaming costs in terms of R and D, investment, and operating costs.
2. Discuss some of the difficulties in the costing of gaming exercises.
3. Define opportunity costs and joint costs. Discuss how these are relevant to the costing of gaming.
4. Discuss how the costs of running the same game at two different locations might differ.
5. What criteria would you use to judge whether a game has been adequately documented? (Specify what you mean be "adequately.")
6. Upon what factors should the decision to select between constructing a computerized or a noncomputerized game be made?

5 Facilities

GAMING AND FACILITY REQUIREMENTS

In the last few years there has been a trend toward the construction of laboratories for gaming of many different varieties. Some examples are provided by the behavioral science laboratories at Purdue's Krannert School,[1] the laboratory at the Center for Research in Management Science at Berkeley,[2] the Leuven Laboratory for Experimental Social Psychology,[3] the Polis laboratory at Santa Barbara,[4] the Economics Laboratory at Santa Barbara, the Computer Based Laboratory at the Systems Development Corporation,[5] the NEWS Gaming Facility at the Naval War College,[6] and many others.

The building, space, and equipment needs for games used for teaching, operational, or experimental purposes vary considerably. The replies to a questionnaire concerning the costs of facilities for academic uses of gaming circulated several years ago indicated that the majority of those involved in gaming were utilizing spare capacity in their estimates at no allocated costs.[7]

Games used as supplementary teaching material can be run in the classroom with little extra equipment other than a few forms or other documents if they do not involve computers. Some games (especially the larger business and war games) are played over a considerable length of time, at rates such as a move a day, or a move per week. In such situations, the players either use classrooms for formal meetings or meet informally anywhere and hand in decisions before some specified time. This, in general, requires a computer but not necessarily a laboratory.

When games are played at a rate of a move a day or less, not only do the player facility requirements change considerably—so do the

83

computer requirements. A turn around time of 12 hours or more usually involves no computer-sharing problem. For even the largest of business games with access to large computational capabilities, a few minutes on a large machine is usually more than adequate. This means that decisions can be keypunched or entered in another manner and given to the computer center so that the game can be batch processed without incurring difficulties in scheduling. There is now a trend towards time-sharing. Hence decisions can be entered from consoles at widely dispersed points gathered together in a program at the computer center and run at a convenient time.

In general, much of the use of small games for teaching can be obtained with few if any special facilities. If teaching games become inordinately large, such as the game used at Carnegie Mellon University, facilities are needed for clerical work and for the organization and storage of programs, special subroutines, analysis programs, and modifications. Adequate facilities are a necessity for handling the library of material that quickly builds up around a complex game.

Teaching games used in a manner such that the players complete the game in one session are, for the most part, noncomputer games or make use of time-sharing facilities. They can generally be run in a concentrated two- or four-hour session without the need for other special facilities. However, if it is important that there be no direct interpersonal interaction between the players, then items such as individual booths become important, and the construction of a laboratory may be called for.

If extensive data processing or extensive use of data banks and files is contemplated, then it becomes extremely important to either have a small, dedicated computer or to use time-sharing. A few years ago, another alternative was to tie up a computer for the length of the session; this involved making arrangements for a priority system that could even result in the dumping of incompleted jobs of other individuals using a common computation center. This was extremely difficult to do, and fortunately is now becoming unnecessary.

Most games played in one uninterrupted session tend to have a decision time of 20 minutes or less. Simple games which are frequently used both for experimental purposes and for teaching, such as an iterated Prisoner's Dilemma game, may run as fast as 20 to 60 seconds per iteration. The importance of using time-sharing when the turnaround time for a computerized teaching game is as low as or lower than 20 minutes is stressed when one considers features such as the distance of the participants from the computer, the amount of keypunching to be done, the methods for validating input, the traffic flow of assistants, and the number of forms to be handled. In general, it was and is difficult to stay consistently under a turnaround time of 5 to 10 minutes for any game utilizing a large computer that does not have a direct entry of the decision by the players themselves.

The comments above and the original survey dealt mainly with gaming for teaching purposes at universities. Most of the comments hold true for teaching and training games at war colleges and large corporations or other commercial users. The more formal and routinized the purpose, the more important it becomes to have a laboratory. For example, the World Politics Simulation (WPS), which was based heavily on the Internation Simulation (INS), has been used at the Industrial College of the Armed Forces[8] under the direction of the Simulation and Computer directorate (SIMCOM). The logistics problems caused by running a game of this variety calls for the setting up of at least a temporary laboratory situation. Commercial gaming, such as that done by the American Management Association (AMA) can be carried out using time-sharing and temporary facilities if only one or two runs are contemplated. However, if a game is going to be used frequently, the cost considerations may easily call for a specialized laboratory.

Gaming for educational purposes at the elementary and high-school level calls for other considerations. A good survey has been provided in the book, *Simulation Games in Learning*.[9] There is a considerable literature on computer-assisted instruction and teaching machines.[10] It would require a separate discussion to do justice to this literature. At this time it must be noted that many of the games for elementary teaching are not computerized. Examples of such games are Democracy, Life Career, High School and Community Response.[11] An experiment with the development of games that are computerized for individual instruction by the Board of Cooperational Education Services in North Westchester County (BOCES)[12] is of interest to those concerned with elementary-school education.

Games used for operational purposes, for training, or for experimentation tend to require more formal facilities than does the usual game for teaching. The special apparatus and the creation of realistically simulated environments for large tactical war games or logistics exercises may call for facilities whose costs must be measured in terms of money in millions of dollars. The Navy Electronic Warfare Simulator (NEWS) and the Logistics Research Laboratory at Rand provide examples of costly facilities where a large computer is or was used exclusively over periods ranging from several days to several months in exercises requiring special facilities for the players, as well as observation equipment, such as one-way glass, television monitors, tape recorders, and special data-processing equipment.

Not all of the developments in gaming have indicated growth. The heyday of the commercial gaming laboratory in the aerospace industry was during the mid 1960s. Since that time, many costly gaming laboratories have been transformed into lunchrooms or have undergone other metamorphoses.

When dealing with gaming for operational purposes it is important to distinguish between simulations or completely computerized games and man-machine games. In the military establishment, expenditures running from the tens of millions to the hundreds of millions have been made over the course of the last twenty years on war gaming and simulation for weapons evaluation, force structure studies, and other purposes. The preponderant expenditure has been either for games that are completely computerized or for programs or models that are part of a computerized gaming exercise or simulation. The facilities needed for such activities differ considerably from activities involving live participants.

Much of the work in simulation has been done on a contract basis at operations such as Rand or RAC, Stanford Research Institute, Systems Development Corporation. In these organizations the equipment for simulation consists or consisted of a well-organized computer center combined with the appropriate laboratory and housing for professional personnel. Although a laboratory is not necessarily needed for work with simulations, it must be noted that computer programs even to this day tend to be local products; they do not travel easily. This being the case, the lack of an adequate professional clearing-house system and the lack of documentation are such that there has probably been far less cumulative growth of knowledge than one might have expected for the expenditure.

The two types of gaming for which laboratory facilities are possibly the most critical are man-machine operational gaming, as exemplified by the Rand Logistics Simulation Laboratory[13] and experimental gaming. The considerations that go into the construction of laboratories for these purposes are given below.

A university wishing to have a general-purpose gaming laboratory where experiments in psychology, social psychology, business games, learning, small-group behavior, and so forth, can be run should estimate that it may need to spend somewhere between

$100,000 for a very small facility to $700,000 or $900,000 (as of 1975) for a facility comparable to the laboratory at Berkeley. The major items included in this expenditure will be the building, the laboratory layout, equipment other than the major computer, and the major computer facility.

There is a growing literature on the nature of the building requirements and laboratory layout. An excellent discussion is given in *Progressive Architecture*[14] and in articles by Rijsman,[15] Hoggatt and others.[16] Decisions must be made about display units, input devices, video systems, audio systems, data-processing auxiliary devices, and special measurement devices and their effect on the flexibility of laboratory operations. Other important items tied in both with laboratory design and equipment are soundproofing, ventilation, and removable partitions and other devices that facilitate the rearrangement of available space.

SOME EXAMPLES

A few specific comments concerning laboratories that exist or have existed stress the nature of the requirements discussed above. The behavioral science research laboratory located at a university is typified by the management science laboratory at Berkeley,[17] the Behavioral Science Laboratories at Purdue,[18] or the Leuven Laboratory for Experimental Social Psychology. Leuven is designed to give more stress to small- and large-group study than Berkeley, which stresses the role of interactive computation. The floor space of all three is in the range of 3,000–4,000 square feet.

The Management Science Laboratory at Berkeley

Figure 5.1 shows the floor plan of the Berkeley Management Science Laboratory.

The two principal considerations guiding the design of the Berkeley lab were that the environment should not detract from the experiments and, whenever possible, automatic controls should be substituted for human control over experimental variables.

The Management Science Laboratory makes use of a dedicated computer for time-sharing. There are seven rooms with fixed walls and a large space that can be broken up into cubicles. Extreme care was taken to minimize outside influences (especially acoustic). A survey was taken among the experimentors asking them to describe

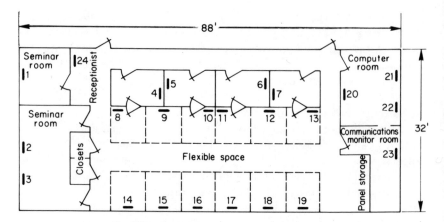

| Standard wall outlets
for all communications

Fixed walls shown as solid lines

FIGURE 5.1

the experiments they had done and were doing and also to make some estimates of what they thought their future load might look like. Space flexibility is achieved by a wall structure with an overhead support so that the floor is not cut up with tracks. This same system has been utilized in the POLIS laboratory at Santa Barbara. A time-shared PDP-5 computer with DECTAPES is used with a memory to link to a PDP-8 computer. The computer maintains control over the experiments and issues commands to the PDP-5 to send information or operate external devices. It is possible to run on 10 teletypes with all input-output in operation without any noticeable delay at the consoles.

The system and the space configuration could be duplicated for about $300,000. Variable costs of operation are relatively low (amounting to a few thousand dollars for maintenance). The capacity of the system is large. Hoggatt and his associates[19] estimate that they could log 280 subject hours in the laboratory in a single day and be able to do the data reduction overnight through the main campus computer facility. The laboratory can also be used in classes and by students in computer courses.

The work of Gerald Shure, both at SDC[20] and at U.C.L.A., is related to that of Hoggatt and his associates but with a somewhat different stress. It has been directed toward the laboratory needs of the experimenter. Shure and Meeker note the reasons for the computer-based laboratory and the needs of the investigator as follows:

We have taken a number of approaches to ease the problem of getting the investigator into the laboratory: (1) developing a "laboratory language" so that the investigator can do his programming; (2) constructing a "universal" laboratory vehicle to a menu of preprogrammed functions; and (3) providing the investigator with a general, interactive vehicle that permits him to rapidly operationalize experimental design logic and poses formulation problems in concrete form, thus giving him the opportunity to "bread board" his experiment by putting him immediately into the operational context and allowing him, by trial and error procedures, to successively approximate his final research design. These quick approaches to laboratory use should significantly reduce the design to implementation phase of experimentation development."[20]

The POLIS Laboratory, U.C.S.B.

The POLIS laboratory at Santa Barbara under the direction of Robert Noel is directed more toward teaching as well as research and to political science, international relations, and sociology.[21] The laboratory is shared by the political science and sociology departments. In contrast with the Berkeley laboratory, face-to-face communication is stressed in the POLIS laboratory. Face-to-face interaction is observed by means of a closed-circuit television system. Video and audio signals are received at a control console in the control center that includes a monitor for cameras and a special switching device for selecting video and audio sources for recording on one or two video tape recorders.

In the type of gaming exercises run in this laboratory, telephone communication is important, as are written messages, reports, and newspapers. Realistic bureaucratic gaming may produce mountains of paper. Thus, reproduction is an important function in a laboratory of this variety. Noel notes that the cost of producing copies is an extremely important consideration in this type of laboratory. For some games, communication via interlink teletypes and a digital computer is used.

There are four different ways in which a computer is used in the support of gaming exercises. They are environmental response to simulation, participant support, data collection, and communications-network monitoring and control. All four uses are made at the POLIS lab. Access to the IBM 360, Model 75, for batch processing is available. Owing to the considerable changes in time sharing it was decided to utilize an external system operated by the Allen Babcock Company rather than have the laboratory locked in to its own time-sharing system which might become outdated quickly.

A certain amount of success has already been encountered with the use of a POLIS network for the playing of international relations games at several different campuses simultaneously, utilizing a time-shared system.[22]

A general description of the POLIS lab is as follows:

Building:

POLIS Lab proper: 3,500 sq. ft.
Administrative control by Department of Political Science; shared utilization with Department of Sociology.

Extended POLIS Lab: 1,000 sq. ft.
Includes the Sociology small groups lab, in which the POLIS Lab has shared utilization rights. All electronics equipment systems are integrated under those of the POLIS Lab, although the small groups lab can be operated autonomously.

Adjacent general purpose classroom: 550 sq. ft.
May be scheduled as part of POLIS Lab. Under-the-floor electronics circuits integrating room into POLIS Lab systems.

Adjacent Sociology Computation Lab: 1,600 sq. ft.
Facility serves as an interface between social science students and faculty and the UCSB Computer Center. Staffed by consultants; maintains a technical library; equipped with assorted desk calculators and three acoustically coupled computer terminals for access to off-campus systems.

Approximate cost of building: $50.00 per sq. ft.

Equipment:

computer, terminals, and specialized hardware	$45,000
video equipment and circuitry	55,000
audio equipment and circuitry	40,000
telephone system (internal)	6,000
trackwall system	50,000
furnishings, etc.	10,000

Operating Budget:

The operating budget which must cover all equipment maintenance, spare parts, programming help, general lab assistance, clerical assistance, and even janitorial service is around $15,000.

Staff:

The regular staff includes one Teaching Assistant and one electronics technician.

Planning and Construction:

The POLIS Laboratory was planned and implemented from the ground up under the direct supervision of its director, Robert C. Noël. Planning began in 1965–66. The building was occupied in March, 1969. The planning, design, fundraising, etc., tasks took at least one-third of Noël's time from inception to completion.

Utilization of the POLIS Lab:

The POLIS Lab is used in a variety of ways. There are laboratory sections of formal course offerings in political science: there are independent studies projects at both the graduate and undergraduate levels; Noël offers a course on social gaming and a course on computer applications in political science, both of which use laboratory facilities; a variety of students use the video and the computer equipment on an informal basis; two social psychological experiments have been conducted in the lab as part of sociology Ph.D. theses; special conferences utilizing gaming have been held in the lab; and, beginning in 1972–73, six new 2-unit "lab/seminars" are being offered to political science undergraduates, which are self-contained learning packages employing simulation and/or gaming exercises and collateral readings.

Topics touched upon in the various lab-related instructional activities include: legislative processes, urban development and urban politics, Presidential elections, foreign policy problems, international relations theory, air pollution, global ecological problems, and social theory. Some of the exercises used are packaged products developed elsewhere; some of them are home grown. Increasingly, a number of students are using the POLIS computing facilities to learn both higher-level language programming and system design and development.

Figure 5.2 shows the POLIS hardware configuration.

The Naval Electronic Warfare Simulator

War gaming can be for both teaching or training and for operational purposes. The early war games required little in the way of formal facilities. Manual war gaming in its various forms has taken place on large-scale maps, on sand tables, or on the floors of large rooms with hexagonal paving. As these games become more elaborate, more equipment and formal facilities are called for. Perhaps the most elaborate facility ever designed is the Navy Electronic Warfare Simulator (NEWS) at the United States Naval War College at Newport, Rhode Island.[23] It is a two-sided electromechanical war gaming system. The NEWS system complete, with automatic damage computation, was operational in 1958 and was modified in 1962. The construction included not only new equipment but also additional player facilities known as "command headquarters." With the modifications, the facility cost approximately $10 million.

The war-gaming department of the college operates this facility. The director heads two branches—an operations evaluation and research division and an engineering and maintenance division. The

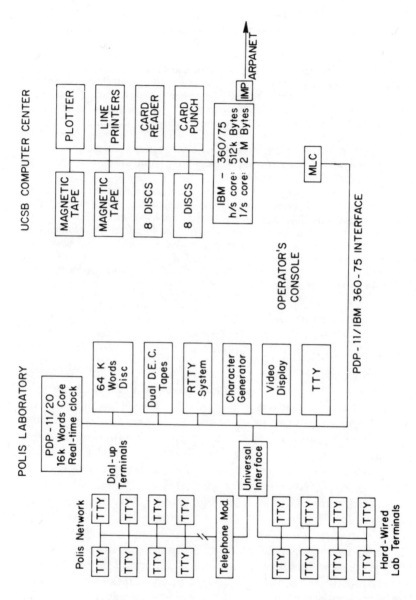

FIGURE 5.2

former consists of 14 officers and 2 civilians while the latter contains 2 engineers and 4 electronic technicians.

The Naval Electronic Warfare Simulator was designed for the play of a two-sided, one-map game at speeds of one, two, and four times real time. The system has three major subsystems: maneuver and display, weapon and damage computation, and communication.

The facility occupies three floors of a large building with 20 player rooms known as command centers on the third floor. There is a larger player room on the second floor, known as command headquarters, and a similar player room on the first floor for the opposing sides.

The control group area consists of an umpire area, a communication room, a control room, and an equipment room. The entire area is dominated by a large master-plot screen covering the area of operations upon which the moving images of all active forces are projected. There is a balcony above the umpire area with seating arrangements for approximately 90 individuals. This area is also used for briefings and debriefings.

The most commonly utilized areas of operation for the exercises are 40 X 40, 400 X 400, and 4,000 X 4,000 nautical miles. Up to 48 active forces can be maneuvered by the players. The forces can be a single ship or aircraft or be aggregations of units. Each force appears on the master plot as a single image. SENSOR systems can also be simulated. In addition to the 48 movable forces there are 14 forces known as fixed forces or fixed targets. These are displayed by hand applied symbols. Cameras are used to take photos of the master-plot screen during play, so that they might be used during the critique. Weather and other such information is also displayed during play. Game clocks and game speed-indicator units are located in the umpire area. A damage computer system calculates the effects of weapons employment. It is a special purpose time-shared on-line analog computer. The NEWS gaming capabilities are also supplemented with manual gaming techniques.

The facility is used for two basic types of games—curriculum games and fleet games. The former are educational and the latter are more operational and analytical in nature. Several types of curriculum games are played. They are the unit-level game, the task-group or task-force level game, and the strategic or national level game. These games vary in length from two to four or five days. The fleet games may be classified into two categories—local play and remote play. The local play games are played at the facility. In remote play games, the players are stationed in their own command

and control locations. The games are generally one-sided with the opposition planned and executed by the officers of the War Gaming Department. In the period 1958 and 1965 approximately 70 curriculum games and 35 fleet games were played.

The NEWS simulator represents one of the largest electronic-tube analog simulators ever built. It is somewhat obsolete and it is being replaced by a digital computer; however, merely from the viewpoint of technology this machine probably deserves a home in the Smithsonian Institute.

The Rand Logistics Simulation Lab

An extremely different type of laboratory is characterized by the Rand Logistics Simulation Laboratory, which was used to carry out investigations of alternative organization structures. The stress in this laboratory was on systems analysis. The nature of the work could be described as demonstration, prototype, and developmental simulation. The techniques employed at the Rand Logistics Laboratory were partially similar to the training simulation techniques used at the Rand Systems Research Laboratory where the gaming simulation stressed the role of organization.

The prime goal of the laboratory was to conduct experiments to discover how the proposed changes in the Air Force logistics systems would work. The staff consisted of psychologists, economists, engineers, mathematicians, systems designers, logistics specialists, and clerical aids. Air Force personnel were used as participants after the laboratory crew had run the preliminary computer models and programs. In the earlier runs, little use was made of experimental design; however, this changed with the usage of the laboratory.

The first major experiment, LP1, was used (1) to test supply policies using selective management, centralized control of assets, and data processing centers, (2) to test supply policies for high-value items using deferred provisioning, responsive procurement, and responsive repair; (3) to test supply policies for low-cost items—using tradeoffs between inventory costs, resupply costs, and depletion costs—automatic resupply, and centralized calculation of inventory levels. Three years of activity were simulated with each year taking about a month of laboratory time. It appears that LP1 had considerable influence on the direction of the Air Force supply system. LP2 was used to investigate what could be learned about the likely support and operations support aspect for new missile systems operations. There was no real world basis in experience for the study. LP3 was used to contrast two extremely different management systems.

The outputs from these simulations would be of value in determining logistics support and operational policies, management control policies, resource requirements, and future systems configurations. A crude overall description of the laboratory facility which took three years to evolve is as follows:

The laboratory floor space was of approximately 2,700 square feet with an additional area of around 200 square feet in close proximity. The laboratory main floor contained 12 microphone input jacks, 100 duplex communication wires for talk paths or signaling circuit service, 6 area loudspeakers, 2 laboratory time period clocks—1 in each of the north and south laboratory areas. There were 10 two-way intercommunication accessory jacks and 25 tons of air-conditioning system. Two-way glass panels in the observation gallery provided for an unobstructed view of activity areas from a seated position at any one of 6 monitored console desks. Carpeting, double flooring and walls, and acoustic tile contributed to insure a sound-deadened area conducive to observation and briefing activities.

A rough approximation of the cost of the construction of a laboratory of this variety and the operating costs of an experiment of the order of LP2 is as follows. Estimates for an in-house capability to perform an LP2 systems analysis in terms of 1960 prices and technology were given:

1. A large-scale high-speed data processing system, on the order of a 7090; price somewhere between $2,500,000 to $3,500,000. This cost would currently be much-reduced.
2. A building suitable for the computer and for office space for an eventual staff of 250 persons with conference rooms and laboratory rooms. The cost was estimated in the neighborhood of $1,500,000 to $2,500,000. Currently this is probably low.
3. An outset staff of 60 persons building up to somewhere between 230 to 270 individuals. The breakdown was: 84–99 professionals; 40–36 programmers and machine operators; 114–135 clerks and secretaries. With this staff one might be able to do a creditable job of analysis and simulation of from 8 to 9 large weapons systems simultaneously.

On the actual LP2 experiment approximately 87 man years of effort and $1,500,000 were expended. The average number of individuals involved was as follows: 10–20 professionals, 3 programmers and machine operators, 10–22 clerical, 3 consultants, 6 participants. An average of 20 professionals was required during the middle half of the project—10 at the beginning and 10 at the end. An average of 22 clerks was required during the laboratory running period. A total of 310 hours of large-scale computer time (IBM 704) and approximately 1000 hours of data-processing time were utilized. The work covered a span of approximately 27 months, including the numerous briefings of various agencies. The suggested phasing requirements and professional backgrounds are indicated for an experiment of the LP2 type.

Phase	Personnel Requirements per Phase
Design	6 professionals
Background Research	6 professionals + 5 clerks + 2 programmers
Modeling and inputs	11 professionals + 8 clerks + 2 programmers
Simulation construction	6 professionals + 15 clerks + 5 programmers
Simulation operation	6 professionals + 25 clerks + 2 programmers
Analysis	11 professionals + 15 clerks + 1 programmer.

The total professional personnel requirements are: three economists, two psychologists, two mathematicians, two engineers, one information specialist, one mathematical programmer, plus five environmental specialists. Computer rentals and services are around $150,000 a year; overhead costs are around $100,000 a year. The schedule for run, design and background research is 0–6 months; inputs, 3–7; modeling, 4–8; simulation construction, 7–11; simulation operational analysis, 11 up.

LABORATORY NEEDS AND THE FUTURE

There is little doubt that the technology of time-sharing is still in great flux, and that the potential role of time sharing in teaching operations and experimentation is enormous. Many of the computational problems that made gaming virtually impossible some years ago are disappearing with the changes in technology. However, as some of these problems become less important, certain basic organizational difficulties that are central to the methods of research, teaching, or operational study come to the fore. In particular, as computation becomes easier, it becomes more important to be able to answer the basic questions of what one wants to do and how one intends to do it now that the computational difficulties do not prevent doing it. The need for the laboratories is a need for organization and coordination, as well as a mere need for space and special equipment. It is easy to gather tons of unprocessed data. It is also extremely easy to perform one's own work in isolation, with insufficient communication with others doing similar work, thus preventing comparisons.

The growth of laboratories signals the needed growth in regional establishments that can absorb the enormous overheads involved in maintaining highly specialized equipment, large data files, and facilities of the variety that make the running of games and the comparison of results and data processing feasible at an economic cost.

In Chapter 4 costs were discussed. It is of interest to note that in all of the publications searched on gaming laboratories, only two had open references to the costs involved.[24] Logical feasibility, technological feasibility, and economic feasibility are different subjects. In our desire to be "scientific" it is easy to discuss the first and possibly the second without paying attention to the third. In reality, attention must be paid to all three. Although cost information may not fascinate the casual reader, in terms of both time and money it is extremely useful to those who contemplate building their own laboratories, running their own experiments, or utilizing gaming for teaching or operational purposes.

GAME CONSTRUCTION AND LABORATORIES

The design of games and the design of gaming laboratories cannot be completely separated from each other. The type of laboratory needed will be determined by the type of games to be run. Researchers with a well-defined area of interest will tend to design their laboratories around extremely specific games. Frequently, the initiating force behind the construction of a laboratory will be an individual with a specific game. For example, much of the work of Rapoport has been directed towards the study of behavior of players in repeated 2 X 2 matrix games. Research work by Shubik, Wolf, and others has been with a specific business game with quadratic payoff functions and with variants of this game.[25]

Those interested in the construction of general-purpose laboratories must make compromises in their designs in order to accommodate the construction specifications called for by their users. A middle-sized general gaming laboratory takes from one to three years to construct and costs in the range of $100,000 to $500,000. The basic design of a good game done by an expert can be sketched in anywhere between a few days and six or seven weeks. The implementation of a game through the stages of programming and debugging until it can actually be used may take anywhere from several weeks to several years. The length of time will depend heavily upon the ease of use of a laboratory setup, as well as the understanding of the individual building the program.

The costs of game construction for operational and experimental purposes are usually much larger than those for teaching. For large war games and logistics exercises, especially, it is difficult to separate the game design, facility setup, and analysis design costs. Informal gaming may usually be done for little money. Simple, well-designed gaming experiments may be performed in a time range of a few weeks to a few months, at a cost ranging from a few hundred dollars to several thousand dollars. The serious analysis of an experiment is another matter. A laboratory is not only a place to run experiments; it is, or should be, a system designed to help the formulation of the experiment to facilitate the running and to automate much of the data processing.

As the use of gaming for virtually all purposes matures, we may expect that the amount of time and the relative costs of the analysis of results and the interpretation of play and debriefing will rise. This will necessarily be accompanied by a further emphasis on the use of laboratories as data processing and organizing systems.

REFERENCES

1. Fromkin, H.L., "The Behavioral Science Laboratories at Purdue's Krannert School," *Administrative Science Quarterly*, *14*, 2 (June 1969), 171–177.
2. Hoggatt, A.C., Esherick, J., and Wheeler, J.T., "A Laboratory to Facilitate Computer-Controlled Behavioral Experiments," *Administrative Science Quarterly*, *14*, 2 (June 1969), 202–207.
3. Rijsman, J., "The Leuven Laboratory for Experimental Social Psychology," *Administrative Science Quarterly*, *14*, 2 (June 1969), 254–259.
4. Noël, R., "The POLIS Laboratory," *The American Behavioral Scientist*, *12*, (July-August 1969), 30-35.

5. Shure, G.H., and Meeker, R.J., "A Computer-Based Experimental Laboratory," *Administrative Science Quarterly*, *14*, 2 (June 1969), 286–293.

6. McHugh, F.J., *Fundamentals of War Gaming*, Newport: U.S. Naval War College (3rd. ed.), 1966.

7. Shubik, M., "Gaming: Costs and Facilities," *Management Science*, *14*, No. 11 (July 1968), 629–660.

8. *World Politics Simulation*, Washington, D.C.: Industrial College of the Armed Forces, 1970.

9. Boocock, S.J., and Schild, F.O., *Simulation Games in Learning*, Beverly Hills: Sage Publications, 1968.

10. Schramm, W. (Ed.), *Four Case Studies of Programmed Instruction*, New York: Fund for the Advancement of Education, 1964. *See also* Schramm, W., *The Research on Programmed Instruction: An Annotated Bibliography*, Washington, D.C.: U.S. Office of Education (OE-34034), 1964.

11. Academic Associates, Baltimore, Md.

12. Wing, R.L., *The Production and Evaluation of Three Computer-Based Economics Games for the Sixth Grade*, Westchester County, N.Y.: Board of Cooperative Educational Services, 1967.

13. Geisler, M., "The Simulation of a Large Scale Military Activity," *Management Science*, *5*, 4 (1959), 359–369.

    Geisler, M.A., and Ginsberg, A.J., "Man-Machine Simulation Experience," Santa Monica, Calif.: Rand Corporation, P-2314, August 1965.

14. "New Tools: Education's Emerging Technology," *Progressive Architecture*, *69*, 4 (April 1968), 134–145.

15. Rijsman, J., *op. cit.*

16. Hoggatt, A.C., Esherick, J., and Wheeler, J.T., *op. cit.*

17. *Ibid.*

18. Fromkin, H.L., *op. cit.*

19. Hoggatt, A.C., Esherick, J., and Wheeler, J.T., *op. cit.*

20. Shure, G.H. and Meeker, R.J., *op. cit.*

21. Noël, R., *op. cit.*

22. Ibid.

23. See McHugh, F.J., *op. cit.*, Chapter 5.

24. Hoggatt, A.C., Esherick, J., and Wheeler, J.T., *op. cit.*

25. Shubik, M., and Wolf, G., *Gaming for Teaching and Research: User's Manual*, Part I: "The Basic Structure of the Market Game for Teaching and Research Purposes," New Haven: Yale University, Department of Administrative Sciences, June 1972.

6 Intention, Specification, Control, and Validation

The taxonomy presented in Chapter 2 provides a first-order description of the purposes behind the use of different types of gaming. The observation was made that several different interested parties might be involved in the same gaming exercise. In experimental gaming, the numbers may be as low as one or two. In general, the experimentalist will have built his own game and might also run it. If he uses live players, the goals of both the experimentalist and of the players must be taken into account. If he has a game which is entirely simulated, the experimentalist can have tight control over the goals of the artificial players.

At the other extreme, when a man-machine game is built by a contractor for a sponsor in a government agency, there may be as many as five or six independent parties involved, each having his own perceptions of what is called for, and all having imperfect communication with the others. For example, a gaming study might be instigated initially by a technically trained officer whose term of office is temporary in the military. He may be sufficiently persuasive to convince his superiors to fund such a project. They or his agency become the sponsors. If the agency does not have technical knowledge itself to have the game built "in-house," a consulting group might be called in. This can result in the introduction of two further sets of participants in the construction of a game. The negotiators and the model-builders or technical personnel are not necessarily the same. In theory, one would expect that the intermediaries would have no influence on the interpretation of the purpose for the project and on the specification of the game. In reality; this is not necessarily so. At the very best, the introduction of an extra bureaucratic step increases the probability for misunderstanding of the original intent. After the game has been built by a contracting group, it may then be delivered for playing and testing to either the instigator or to yet another party who becomes the caretaker or operator of a study that he did not order.

In most academic or research applications of gaming, the bureaucratic structure and the sociological aspects of the decision process involved in sponsoring, building, and using a game are of secondary importance. This is not the case when the games are used for operational purposes and are produced within large bureaucracies. Even in an academic context, it is probably wise to make sure that the purposes of all concerned parties are understood. Furthermore, it is important to check to see that if there are unstated purposes they should be brought into the open or otherwise accounted for in advance. For instance, a study may be carried out for one purpose and fail dismally by any other criterion of measurement. However, after the study has been completed, a new purpose may appear and after the fact it may be shown that the study fulfilled this purpose admirably. This is the equivalent of failing to call one's shots in billiards.

Although the unstated purposes for gaming are possibly the most likely when games are used for operational purposes (for example, a major reason for gaming may be to prevent a decision from being made), there may even be some unstated objectives in academic work. Examples are the following of fads or the utilization of research money that would not otherwise be available under a different title. Possibly, however, in many categories of gaming the greatest unstated purpose is that the individual in his enthusiasm to game or simulate to answer too many questions may have a poor enough specification of his intentions that he actually has no clear purpose.

INTENT

Leaving aside the bureaucratic or sociological aspects of gaming, it is still imperative that we ask of any gaming exercise, "For what purpose and for whom is this exercise being run?" Around 30 different purposes for gaming have been suggested. The intentions of the different users and sponsors differ considerably, and the measures of success are many.

The size of the box office receipts is a good criterion for evaluating the success of a spectator sport from the viewpoint of the promoter. The number of people doing crossword puzzles is a good

101

criterion for the owner of a newspaper. The criteria applied by a general, a zero-sum game-theorist, and a military hardware expert to force posture and allocation games are different from the criteria that might be used to judge the success of a political-military exercise run by a mixed group of political scientists and top government officials. This, in turn, would be different from the judgments applied to evaluate the worth of a teaching game designed to give high-school students an appreciation of international relations.

In spite of the diversity of intent in the use of games or simulation, there is a procedure that can be applied to the construction and use of any game or simulation. It can be described in terms of a set of the following relatively straightforward questions:

1. Who is involved in the game construction?
2. What are the goals of each of the involved parties?
3. What are the priorities of their goals?
4. Is it possible to describe measures for the goals?
5. How do resource constraints in terms of costs and time influence these goals?

These same five questions apply not only to the construction of the game or simulation but to its use as well. Obviously, when large-scale models are constructed, the intention of their sponsors is to have them used. However, added complications appear in use that may easily be overlooked. For example, the construction of a game may not involve the use of the players who must operate the game. It is easy to overlook the effect of the knowledge, intentions, and abilities of these players. Thus, one might construct an otherwise excellent model whose use is vitiated by the failure to account for the goals or understanding of the players. This is especially true in attempts to interpret the play of simple matrix games. It is also true when the role-playing requirements and the motivation of the players in operational or research games are not appropriately controlled.

Assuming that all involved parties have been identified and all are willing to specify their goals, we are ready for the third question. It is not possible to know in advance what to expect from a new venture. Often good experimentalists may run a game with a specific hypothesis to be tested. After they have finished, they discover that it was not possible to test this hypothesis. However, the experience gleaned from the construction and running of the game was worth the effort. The wise game constructor frequently has several goals in mind, so that his efforts are not completely wasted if he fails to achieve some of these goals. Success or failure, however, is not purely random. It is heavily influenced by the establishing of priorities among the goals before the construction of a game.

The specification and statement of goals is important, but it is not enough. The way in which they are to be measured must be noted. This enables one to make sure that one recognizes what one is looking for upon encountering it. The task of constructing criteria and measures precedes that of data gathering and statistical analysis. It is also economical to perform tasks in that order. It is possible to reject a scheme on the basis that it is logically inconsistent or ill-defined without having to go to the expense of first gathering large amounts of irrelevant data.

In some well-defined simulations, for example, those dealing with queuing and delays in logistical systems, it is frequently not too difficult to specify all of the measures, such as delay times, length of queues, size of inventories, and so forth. The importance of these items is reflected in the payoff function or loss function attached to the whole exercise. Thus, for example, the goal of a logistics exercise might be to cut down the level of inventories in a system, given a certain level of performance. The objectives of a game for teaching high-school students may not be so easy to specify. The constructor may say that the purpose is to provide the students with a better education than they are currently receiving. However, the problems entailed in operationalizing such a statement are enormous. What are the dimensions for measuring better education? There is the danger that the amount of enjoyment evinced by the participants is confused with the amount of education they may have received.

Even though the intentions of the game constructors are known, there still remains the task of costing to be done before we can describe the payoff to running a game or simulation. The construction and running of a model are not free goods. The worth of the exercise must invariably be compared with its costs in terms of both money and time. There are many techniques which might be technologically feasible but must be ruled out when costs are considered. Before constructing an elaborate model, it is necessary not only to find out what the goals happen to be but also to estimate what we are willing to pay to achieve them.

SPECIFICATION

Many of the remarks made here clearly refer to the problems faced in modeling and experimentation in general, not gaming and simulation alone. It would be presumptuous to claim any particular originality in these comments. Nevertheless, it is important that they be made and be made frequently. It is with this thought in mind that these observations are made, even if they contain little new knowledge for the skilled model-builder. Almost all of the literature on the validation of simulations and on gaming presents a mixture of discussion on specification, control, and validation with the emphasis on validation. In most scholarly articles, it is implicitly assumed that the intent of any study is clear.[1]

The task of specification is that of building the appropriate model. Undoubtedly, model-building is still an art form. However, there are a host of specific valuable rules which can be applied to the construction of virtually any game or simulation. Furthermore, these rules should be applied well in advance of any large-scale data-gathering and statistical processing. There are four general rules which apply to model building in general:

1. Evaluate and be prepared to borrow or modify the work of others.
2. Sketch out the model and examine it in terms of relevance.
3. Examine the model in terms of consistency and closure.
4. Do not assume anything that is not needed.

A certain amount of duplication of effort is desirable as a means for checking of one's own work and the work of others. However, especially where games and simulations and their results are to be used by a community, it is highly desirable to take advantage of the work of others. Even though the state of documentation is currently poor, there is usually enough informal communication to enable one to check to find out if a model closely resembling the one in mind has already been built and used. It may turn out to be cheaper to build a new model rather than to modify the other. However, the gain in having seen it may be large.

Relevancy is a key to good model-building. Irrelevant realism is one of its most potent enemies. Relevancy calls for an intermixture of empirical knowledge, understanding of what the intent of the program is, and an ability to abstract.

The need to check a model for consistency and completeness is obvious. However, it is worth noting that especially when models or scenarios are complex and not strictly mathematical or symbolic in structure, it is relatively difficult to work out an adequate easy-to-apply method for checking consistency and completeness. One of the important features in using simulations is that the construction of a well-defined simulation provides the method to check for completeness and consistency of a large model. In some instances the goal or intent of the builder may seem to be no more than to check the consistency and completeness of a model without even being particularly concerned about the "validity" of the numbers he can generate with that model.

Not assuming anything that is not needed is the last but by no means the least important of these four general rules. In many gaming and simulation exercises, there is a tendency for the models to grow large quickly. Extreme care must be manifested in cutting them down to size. This is closely related to the nature of relevance, but it is not the same. A complex model can well be relevant. However, it might be possible to simplify it considerably without losing relevancy.

The ten more detailed checks to be applied to any game or simulation are as follows:

1. dimension checking
2. specification of measures
3. the selection of simple functional forms
4. the minimization of the number of parameters
5. aggregation
6. symmetry
7. continuity
8. limiting behavior
9. sensitivity
10. direction

Dimension checking and the specification of measurements have already been noted in reference to the specification of goals.[2] These techniques are relevant to any model-building effort.

Select relationships and functional forms to be as simple as possible. If it yields a good enough approximation, use a linear relationship. Otherwise, contemplate piecewise linear relationships or quadratics. There is an art to parsimonious description and it is an art that is important to practice.

Closely related to the selection of simple relationships is the effort to minimize the number of parameters in the model. Those who call for realism in modeling may produce a model that has so many degrees of freedom that virtually nothing can be established about its behavior.

Selecting the correct level of aggregation in a model is almost always important and frequently difficult. The behavior of an aggregate unit, especially in a game, is by no means the same as the

105

average behavior of the disaggregate. In our discussion of a game-theory background for gaming, part of Chapter 1 was devoted to the discussion of solutions. Solutions represent the results of the behavior of players. Solutions do not change in a simple manner as players are aggregated. A major failure in political, economic, and social models is brought about by selecting the wrong level of aggregation. The expense of disaggregating is usually large and must be weighed against obscuring the phenomena one is trying to study.

The process of aggregation in games and simulation is neither smooth nor simple. One must determine whether there are good empirical reasons to expect that the system should exhibit a smooth behavior under aggregation. If there are, these may provide a justification for constructing an aggregate model with such properties. If there are not, the dangers of aggregation must be faced up to and a sensitivity analysis is called for.

Symmetry is always important in model-building. It is closely related to aggregation but different. For example, it may be observed that all members of a class of individuals are virtually identical. Given that a symmetrical representation of all members is possible, the modeler may be tempted to aggregate the class and to replace it with a representative individual. The dangers of aggregation still apply. It may be much simpler to operate with a class of individuals, each one of whom is a replica of the other. Yet it may not be sound to try to replace the group with a composite player. Every one of 100 voters in a society having 100 voters might be regarded as identical. However, an attempt to replace 51 of the voters by a composite voter with 51 votes would be disastrous for most purposes.

We have already noted the distinction, so important in game theory, between external symmetry and internal symmetry. External symmetry implies that all conditions that have not been explicitly specified within the model apply symmetrically to all. Internal symmetry refers to symmetries that have been explicitly specified within the model. Game theorists and mathematical model-builders in general are frequently concerned with the behavior of players whose specific properties are the same, but who are playing in nonsymmetric games. Much of the work in behavioral science has been devoted to individuals who have different properties playing in symmetric games. The use of symmetry often enables large simplifications to be made in systems where aggregation is not possible. It is much easier to study mass societies or markets or polities in which the individual units are assumed to be identical in their basic properties although they are not aggregated than it is to study more highly aggregated systems with different individuals.

Does the game or simulation manifest the appropriate continuity of behavior? In general, we know something about the stability or the lack of stability of the population we are trying to model. It is possible to check the outputs of a model or subsections of a model to see if the appropriate continuity in behavior is observed. If it is not, it may be that the model has served to illustrate an instability in the real world. It is most probable, however, that the discontinuity was an artifact of the model. It might have been caused, for example, by a boundary condition that does not exist in the real world, or by the change in functional form in the approximations used to represent behavior. A good example is provided trying to model the inventories of food in a restaurant. In the real world these inventories and the smoothness of their change depend heavily on the skills of the chef as an imaginative chef rather than on the operator of an optimum-inventory reorder policy.

In some societal and military models, limiting behavior is important. For example, when considering battles of tanks or aircraft on bombing missions, one might want to know if a system reaches a saturation point. Furthermore, one might want to determine when the increase in the number of individuals in an economic, social, or political context gives rise to a basic changeover from individual to mass behavior. It is important to be able to check to see that the effect of an increase of numbers appears naturally and correctly in a model. Frequently, fundamental difficulties that were overlooked are discovered when an attempt is made to check for limiting behavior. Recent work in economic theory provides many examples of new phenomena appearing when limiting-behavior processes are defined. For example, if a system has more and more different types of goods, there are several different ways in which the relationship among these goods can be modeled. Do substitutes crowd together as the number of the goods in the economy increases, or do they remain apart?

A key factor in virtually all gaming and simulation is sensitivity analysis. This refers to how the system reacts as a whole to a change in the value of an initial condition or some other parameter. When a system contains hundreds or even thousands of parameters it is obvious that a systematic sensitivity analysis for each parameter and for all combinations of parameters is out of the question. The problem is not primarily a mathematical or a statistical one. It rests far more with "knowing your business."

If the model-builder builds an extremely complex model of a specific aspect of human affairs, he is presumed to know something about the problem. If this is the case, then it is not unreasonable to expect that for any particular question being asked of a model, an expert should be able to specify the most important parameters. Furthermore, he should be able to suggest what he expects might happen as a result of varying various combinations of these parameters. Thus, one should be able to use the expert to select a few out of the infinite combinations of parametric changes in order to examine the sensitivity of the system. Such a sensitivity analysis will lead to one of three conclusions. (1) The system will behave more or less in the manner predicted by the expert. This means either that the simulation or game is good or that everybody has succeeded in building their prejudices and misconceptions into the model in such a way that it reflects their rationalizations. (2) The system gives counter-intuitive results, in which case the designers may learn something new. (3) The manifestation of counter-intuitive behavior may indicate that the model is no good.

It has been suggested that the expert should try to guess what he expects is going to happen. It is not unreasonable in certain key instances to ask the expert to guess the direction that will be taken by some of the variables in the system. Frequently there are good empirical or theoretical reasons to support predictions of the way in which the system or parts of the system should behave. If the expert both makes a guess of what he expects and specifies in advance why he expects it, then the chances that something will be learned from the run are relatively high.

The simple study of the combinatorics of parameter setting when there are even a few dozen parameters, each of which can only take two or three values, should be sufficient to convince anyone that an exhaustive search of the behavior of a model by mathematical and statistical means, without making detailed use of expertise, specific knowledge, and a preliminary understanding of the details built into the model, is the wrong and uneconomic approach.

CONTROL

In order to best describe control we must separate the game into two parts—the simulation or model and the actual game. The comments concerning the model apply to simulations regardless of whether they are also games. Simulations range in complexity and accuracy from a simple model of a specific and easy-to-measure physical process to enormous programs purporting to model the behavior of whole economies or societies. For most individuals interested in gaming, the key to the difficulties in control comes in the size of the system and the role human behavior plays. In general, the smaller the system and the less important human behavior, the easier it is to control the system. Simulations designed primarily to study the flow of goods through machines in a job shop present extremely different control problems than a large-scale international-relations political game, such as INS. In between the two of these are simulations in which, although there may be considerable physical structure, an emphasis has been placed on human factors. An example is given of the simulation of the behavior of a crew of a destroyer.[3] For purposes of classification and communication it is worthwhile to distinguish three types of simulation: tactical technical simulations, human factors simulations, and strategic and behavioral simulations. Most of the work in operations research falls into the first category, human factors experiments into the second, and large-scale political and economic models into the third.

As with simulations, there are many distinctions that can be made among games. For our purposes the two important distinctions that have the greatest implications for control are the differences between a free-form game and a rigid-rule game. A political-military exercise with a control team ruling upon moves provides an example of the former, and a simple, computerized matrix-game experiment provides an example of the latter.

The importance of control in any gaming exercise comes first in its effect upon the model or simulation and then upon the actual running of the game. In the construction of a complex model, one need not separate out the work on specification from that on control, as they will take place simultaneously in the model-building. Nevertheless, specification is designed to make sure that the model reflects the phenomena and stresses the interests of the instigator, whereas control stresses the manageability of the model. This being the case, the following items must be considered in the construction of the model from the viewpoint of control:

1. design of input
2. design of output
3. design of data banks and data gathering
4. design of final data processing

5. documentation
6. design of internal checking procedures
7. specification of the role of random variables.

The value of a simulation varies inversely as the amount of output. Large quantities of output usually indicate that the user does not know what he intends to do with the data. An important feature of control is to stress the design of output in such a way that one is saved from having to throw away large amounts of unused data later. It is possibly a good strategy to worry about output even before considering input.

Another important feature of control is the design of data banks and data-gathering procedures. The availability and form for use of data are technical housekeeping problems that do not appear particularly important on the surface but have considerable influence on the way a simulation or a game will be used. Related examples can be seen when we consider the usage of libraries and computing systems. If services are not made easy for the user or if he is not properly educated, he will be slow to use the system.

The design of data processing is a control function closely connected with, but not the same as, validation. The task being referred to here is more like housekeeping. Individuals who are perfectly good statisticians may not carry out certain tests if the time involved in assembling the data is too great or if running the tests is too costly. The control element in data processing should emphasize the ease and economy with which data can be processed.

Documentation at all levels for a simulation or game is extremely important and is probably in general the most poorly performed control function. Documentation is absolutely vital if the game or simulation is to be used by anyone other than its builders. Even if the use is restricted to the builders it is frequently important to know what was done when. Good documentation is tedious and expensive. Early rigorous control makes both the tedium and the expense less.

If errors can occur they will. (For the cognoscenti this is the simplest statement of Murphy's Law.) Modern computer languages supply a fair amount of internal checking, so that a logically inconsistent program has a fair chance of not running. However, machine-language internal-checking routines are not designed to catch substantive errors.

The role of random variables must be taken into account in the control of a simulation. If it can be shown that a mean value will serve to adequately guide the behavior of the system, then an enormous saving can be made in the length and cost of runs needed to explore the behavior of a simulation. Frequently one cannot replace random variables by mean values, in which case an important design problem in minimizing the costs of experimentation is posed.

The control features of the running of a game involve:

1. design of the pretesting of players
2. design of the briefing of players
3. design of the posttesting of players
4. design of the debriefing of players
5. control of cues and artifacts
6. choreography and laboratory discipline

The design of pretesting and posttesting of players appears to be self evident. However, surprisingly few gaming exercises have used these controls. One reason is probably that the costs of using such controls are high without a highly automated and organized laboratory. Another reason is that individuals working in different disciplines often overlook the contribution that other disciplines can make to their work. Thus, military gamers might assume that the sociological or social-psychological aspects of the players are irrelevant to their results. Economists and operations researchers tend to control for different aspects of the game than do social psychologists or sociologists. For some purposes it might be reasonable to argue that pretesting and posttesting of players is not necessary. In such instances this should be explicitly stated.

The briefing of players is a key element in the control of any gaming exercises. Should briefings take place by means of written instructions, should they be given to all individuals present in a large hall, should there be an open question period between the players and the control or should questions be submitted in written form? These are all questions that still need further research. There is no simple blanket answer. The restrictions on briefings obviously vary with the type of exercise being held.

In many gaming exercises there may be no need for a debriefing other than as a courtesy to the players. In others, such as operational games in which the players will be involved in the operation, the debriefing session is possibly the most important aspect of the game.

One man's experiment is another man's noise. One professional's theory is based upon accepting as given the unknowns that other professionals spend their lives studying. The control of cues and artifacts is important to games for virtually any purpose. Political scientists such as Schelling[4] have suggested that cues play an extremely important part in coordination and signaling between indi-

111

viduals and organizations. Experimental psychologists have stressed the importance of checking the design of experiments for artifacts. Even in running of the simplest games, control should be exercised to check for the influence of phenomena such as the length of moves, the way in which information is presented, termination effects, the optimum noise level under which the experiment should be performed, the role of boredom, and the influence of extraneous random events. Individuals code information, and cues and artifacts are frequently efficient coding devices. Hence, with the development of theories of human information-processing, we may expect that a far deeper understanding of the role of cues will be possible.

Some examples of the control problem are boredom caused by allowing too long a time period for moves so that the brighter or quicker players "switch off" or try to otherwise occupy the slack time after they have moved. There is also the "happy-hour" effect, which may be manifested if a game that is meant to be of indeterminate length is running at two or three in the afternoon on a Friday when the players know that they will not be playing the game during the next week. If the environment in which a game is played is too noiseless and sterile, then items which might otherwise have been overlooked in a higher noise level will serve as cues. For example, if the first move of a game involves nothing more than selecting a number between 1 and 20 and there is no scenario that gives a context for decisions, it is possible that the player will select "magic numbers," such as 3, 7, or 13, that he will select the number at the upper-left or the lower-right corner of a matrix, or that he will select a number based upon counting the number of flies buzzing around the room or the number of drips he has heard fall from a nearby water tap.

The design of good laboratory discipline and the choreography for running a gaming exercise smoothly is not a particularly intellectual exercise and entails a great amount of relatively dull and painstaking work. Yet, at this stage of the development of gaming it appears that at least for some games, such as political military exercises, a good game will not be feasible without a carefully controlled scenario and a good control team.

VALIDATION

Given that the intention for which a model has been constructed is known, the hypotheses to be tested, if any, have been clearly stated, the model has been adequately specified, and sufficient control has been taken that the game portrays what it purports to portray, the problems of validation still remain. The tasks that must be performed in the validation of any simulation or gaming exercise are:

1. reconfirmation of goals concerning intention
2. validation of model internal-logic concerning specification
3. the check of the model as a reflection of theory
4. the check of the dependability of data sources
5. design of experiments to measure parameters or check sub-models
6. design of statistical tests on outputs
7. design of experiments on system behavior
8. design of field tests for comparisons with the simulation

The following items refer more specifically to the operation of a game than to the model or simulation:

1. analysis of player-pretest information
2. checking of player-briefing information
3. analysis of player-posttest information
4. checking of the validity of cue and artifact control
5. analysis of player behavior
6. checking for totally different explanations of behavior

In the busywork associated with the construction of games and simulations, expecially if they are large and involve people from more than one organization, it is easy to forget the original purpose or to modify the purposes during the process of completing the construction of the simulation. The first important steps in validation amount to going back and checking over the previous work. This includes reconfirming the purpose of the model as well as rechecking the construction. In some instances where gaming is being used experimentally to check a theory, the first step in validation is to see how well-suited the game is to this purpose. It is easy to become a victim of the "black box phenomenon" when dealing with large-scale models, especially if they are complicated and poorly documented. One has to take a great amount of work on faith. However, it is always important to do a certain amount of backchecking.

A particularly important area in which rechecking must be done by the validator is examining the dependability of data sources. People feel more comfortable if they are able to reduce the amount

113

of uncertainty in their environment. In any intellectual endeavor, one way to reduce uncertainty is to quote authorities and experts. Without suitable cross-checking this can lead quickly to some extremely dangerous distortions in basic data. For example, if an individual with a good reputation makes extemporaneous remarks during a speech, such as guessing the value for some critical parameter, these remarks can be picked up and reported in the proceedings of a conference. Another professional could read the proceedings of the conference and quote the value of the parameter in a technical report. If the second professional also has a good reputation, the stage is now set for someone to build a simulation and use this parameter quoting as his authority a technical report written by a responsible scientist. A conscientious individual experimenter may avoid this type of booby trap; however, when large-scale simulations or games are built by contractors for agencies the danger of this type of phenomenon appearing is great.

On the assumption that all of the preliminary steps of checking have been carried out, the statistical problems of validation remain. Among the methods that can be used are the analysis of variance, factor analysis, nonparametric tests, regression analysis, and spectral analysis. There is a growing literature on the statistical problems of validation of simulation.[5]

The articles of Fishman and Kiviat provide a background for understanding the importance and the methods of using spectral analysis and the study of time series obtained from simulations; they observe that because simulation data are generally autocorrelated that the statistical tools used for studying independent observations will lead to an overestimation of the reliability of the sample means and variances.[6]

The literature on nonparametric statistics, design of experiment, econometric methods in general, and time-series analysis in particular are all more-or-less directly applicable to the statistical problems faced in validating simulations and games. It is beyond both the scope and the intent of this book to deal with the specifics of any of these. For the interested reader several references are supplied.[7]

The last factor in validation activities concerns checking for totally different explanations. This involves taking a last look backward after all the work has been done to consider the possibility that there might be a completely different set of hypotheses that could explain the behavior even though the hypotheses tested for appear to be borne out. For example, each professional is to some extent a prisoner of the commitment he has to his own profession. Thus, psychiatrists and social psychologists have an extremely different world view of, say, efficiency and conflict than do operations research personnel and game theorists.

A game theorist experimenting with a 2 X 2 matrix knows that there are many alternative theories of conflict or cooperation that may determine the same behavior. A social psychologist may have theories of behavior based on social psychological variables, which are different from those of the game theorist. He may not even be aware of the game-theoretic nuances of his experiment. A completely different set of hypotheses drawn from a different discipline might explain his results as well as, if not better than, his own hypotheses. This problem is at the heart of the difficulties in experimenting with games such as INS. One does not know from the results or from the hypotheses tested whether or not there are dozens of other hypotheses which would explain the observations better or at least as well as those used.[8]

CONCLUDING REMARKS

Mayberry suggested in the course of a panel discussion at which this author was present that "a simulation was valid when both the analyst and the customer, after adequate consideration, were satisfied with it." We might rephrase this as "a happy customer is a validated simulation." If the customer knows what he wants and is relatively competent and the builder of the simulation is honest and professional, this definition is by no means as flip as it may appear to be at first sight. Van Horn,[9] Naylor,[10] Fishman and Kiviat,[11] and others have observed that the word "validation" is used as a blanket word to cover many different problems that must be faced in deciding whether or not the construction of a simulation and/or the running of a game has been worthwhile.

Fishman and Kiviat[1] (as has already been observed by Van Horn[1]) have divided simulation testing into three categories: (1) *verification* insures that a simulation model behaves as the experimenter intends, (2) *validation* tests the agreement between the behavior of the simulation model and a real system, (3) *problem analysis* embraces statistical problems relating to (the analysis of) data generated by computer simulation.

Naylor and Finger[1] have suggested a multistage verification of computer simulation. The first stage calls for the formulation of a set of postulates or hypotheses describing the behavior of the system of interest. The second stage calls for an attempt to verify the postulates on which the model is based, subject to the limitations of

115

existing statistical tests. The third stage consists of testing the model's ability to predict the behavior of the system under study.

Van Horn suggests:

> When adequate data are available, statistical tests are an essential part of validation, but the overall validation process should encompass much more. In rough order of decreasing value/cost ratios, some of the possible validation actions are:
>
> 1. Find models with high face validity.
> 2. Make use of existing research, experience, observation and other available knowledge to supplement models.
> 3. Conduct simple empirical tests of means, variances, and distribution using available data.
> 4. Run "Turing" type tests.
> 5. Apply complex statistical tests on available data.
> 6. Engage in special data collection.
> 7. Run prototype and field tests.
> 8. Implement the results with little or no validation.
>
> The real task of validation is finding an appropriate set of actions.[1]

In this chapter it has been suggested that instead of using the word "validation" to cover several different activities, the roles of intention, specification, and control should be recognized for their own importance. Tied in closely with these is the understanding that the acceptance of a game or simulation as having performed its task to the satisfaction of all concerned cannot be separated from an appropriate understanding of the intent for which it was designed, the costs involved and the alternatives available. Even pure statistical testing cannot be considered without taking into account its costs and the value of the additional accuracy it may provide.

The substantive details would differ among what the alternatives are, what the costs are, and what the goals of gaming or simulation for operational purposes, experimental purposes, teaching purposes, and so forth, are; they would be relative to the specific application at hand. It is maintained, however, that the principal aspects of intention, specification, control, and validation apply in general to all of these purposes.

EXERCISES

1. Distinguish among intention, specification, control, and validation problems in the running of gaming exercises.
2. Give an example in which the aggregation of individual decision units beyond a certain level may distort the behavior of a system.

3. Discuss the uses of symmetry in modeling multi-person conflicts. Provide examples to illustrate its uses and its limitations.
4. What is the difference between external and internal symmetry (or extrinsic and intrinsic symmetry)?
5. Discuss what is meant by sensitivity analysis, and its importance in the checking of games or simulations.

REFERENCES

1. A sample of the literature on validation and allied problems is as follows:

Alker, H.R., and Brunner, R.B., "Simulating International Conflict," *International Studies Quarterly*, Vol. 13, No. 1 (March 1969).

Boocock, S.S., and Schild, E.O., *Simulation Games in Learning*, Beverly Hills: Sage Publications, 1968.

Chadwick, R.W., "An Empirical Test of Five Political Assumptions in an Internation Simulation about National Political Systems," *General Systems*, *12* (1967), 177–192.

Cohen, K.J., and Rhenman, E., "The Role of Management Games in Education and Research," *Management Science*, *7*, 2 (January 1961), 131–166.

Conway, R.W., "Some Tactical Problems in Digital Simulation," *Management Science*, *10*, 1 (October 1963).

Fishman, G.S., and Kiviat, P.J., "The Analysis of Simulation Generated Time Series," *Management Science*, *13*, 7 (March 1967).

Fishman, G.S., and Kiviat, P.J., "The Statistics of Discrete Event Simulation," *Simulation*, April 1968, 185–195.

Hermann, C.F., "Validation Problems in Games and Simulations with Special Reference to Models of International Politics," *Behavioral Science*, *12* (May 1967), 216–231.

McKenney, J.L., "An Evaluation of a Business Game as a Learning Experience, *Journal of Business*, *35*, 3 (July 1962), 278–286.

Mayberry, J.P., "Validity of Simulation Models," Talk presented at Symposium on Computer Simulation as Related to Manpower and Personnel Planning, Annapolis, Md.: U.S. Naval Academy, April 1971.

Naylor, T.H., and Finger, J.M., "Verfication of Computer Simulation Models," *Management Science*, *14*, 2 (October 1967), 92–101.

Naylor, T.H., (Ed.) *The Design of Computer Simulation Experiments*, Durham, N.C.: Duke University Press, 1969.

Patchen, M., "Models of Cooperation and Conflict: A Critical Review," *Journal of Conflict Resolution*, *14*, 3 (September 1970), 389–408.

Rapoport, A., and Orwant, C., "Experimental Games: A Review," *Behavioral Science*, *7*, 1 (January 1962), 1–37.

"Review of Temper Model," Princeton, N.J., *Mathematica*, September 1966.

Van Horn, R.L., "Validation of Simulation Results, *Management Science*, *17*, 5, January 1971.

Wing, R.L., *The Production and Evaluation of Three Computer-Based Economics Games for the Sixth Grade*, Westchester County, N.Y.: Board of Cooperative Educational Services, 1967.

117

Rice, D.B., and Smith, V.L., "Nature: The Experimental Laboratory, and the Credibility of Hypothesis," *Behavioral Science*, 9, 3 (July 1964), 239–246.

2. For a further reference on dimension checking *see* De Jong, F.J., *Dimensional Analysis for Economists*, Amsterdam: North Holland, 1967.

3. Siegel, A.I., and J.J. Wolf, *Man-Machine Simulation Models: Psychological and Performance Interaction*, New York: Wiley, 1969.

4. Schelling, T.C., *The Strategy of Conflict*, Cambridge: Harvard University Press, 1960.

5. *See*

Naylor, T.H., Balintfy, J.L., Burdick, D.S., and Chu, K., *Computer Simulation Techniques*, New York: Wiley, 1966.

Naylor, T.H., and Finger, J.M., *op. cit.*

Naylor, T.H., Wertz, K., and Wonnacott, P., "Some Methods for Analyzing Data Generated by Computer Simulation Experiments," *Communications of the ACM*, 1967.

Fishman, G.S., *Concepts and Methods in Discrete Event Digital Simulation*, New York: Wiley, 1973.

6. Fishman, G.S. and Kiviat, P.J., *opera cit.*

7. Blackman, R.B., and Tukey, J.W., *The Measurement of Power Spectra*, New York: Dover Publications, Inc. 1958.

Clarkson, T.P.E., *Portfolio Selection: A Simulation of Trust Investment*, Englewood Cliffs, N.J.: Prentice-Hall, 1962.

Cochran, W.J., and Cox, G.M., *Experimental Designs*, New York: Wiley, 1957.

Draper, N.R., and Smith, H., *Applied Regression Analysis*, New York: Wiley, 1966.

Fishman, G.S., *Digital Computer Simulation: Input-Output Analysis*, Santa Monica, Calif.: The Rand Corporation, RM-5540-PR, 1968.

Jenkins, G.M., "General Considerations in the Analysis of Spectra," *Technometrics*, 3, 2 (May 1961), 133–166.

Siegel, S., *Nonparametric Statistics*, New York: McGraw-Hill, 1956.

Tocher, K.D., *The Art of Simulation*, Princeton, N.J.: Van Nostrand, 1963.

Walsh, J., *Handbook of Nonparametric Statistics*, Princeton, N.J.: Van Nostrand, 1962.

8. *See* Chadwick, R.W., *op. cit.*

9. Van Horn, R.L., *op. cit.*

10. Naylor, T.H., Balintfy, J.L. et al, *op. cit.*

Naylor, T.H., and Finger, J.M., *op. cit.*

Naylor, T.H., Wertz, K. and Wonnacott, P., *op. cit.*

11. Fishman, G.S. and Kiviat, P.J., *opera cit.*

7 A Guide to Information Sources on Gaming and Related Topics

ON THE CLASSIFICATION AND EVALUATION OF GAMING LITERATURE

There is no substitute for knowing your trade, and there is no joy in spending dozens of semifruitless hours searching through library stacks. The problem of classifying, categorizing, and evaluating professional literature belongs primarily to the professionals. One of the pervading misconceptions about the use of libraries and the abilities of librarians is that somehow or other a good librarian who is not necessarily a professional in any particular subject other than library science will be in a position to determine how to evaluate as well as categorize books and articles and will be in a position to know what should be kept and what should be rejected.

When a subject for study is a new, growing, and somewhat ill-defined discipline, the problems of literature search and evaluation are more complex than the problems faced in literature search in a well-established discipline. Gaming means many things many people and has many levels of application, making it virtually impossible to present the literature so that the scheme will satisfy all the needs of the many different users. Any classification scheme which attempts to do so must be built with a strict recognition of its limitations. At best it is an associative scheme which will help *guide* the reader in searching literature. It should save him considerable time and help him to avoid dead ends. The classification scheme offered here and the literature discussed in the subsequent chapters are offered to assist and hopefully to save time for the reader. The taxonomic scheme used here is obviously not sacrosanct. There are many other different ways for classifying the gaming literature. Furthermore, many other ways will stress different points.

119

Some Notes on Scholarship

Before the classification scheme is discussed, I wish to note five propositions from the mythology of scholarship:

1. An overkill of an article with footnotes is a sign of a high degree of scholarship.
2. The cost in time and money of individual investigation of scientific subjects is not part of true science or scholarship.
3. Science must always be objective. Hence, it is a "no-no" to suggest to one's colleagues and students that certain articles are not worth reading or that other articles are good. By doing so, one is being "subjective."
4. The task of refereeing articles should be a reward in itself for the referee whose name will remain anonymous and whose professional credit for doing the work will be small or nonexistent, when compared with spending that time publishing an article of his own. (This fault could easily be curved by publishing annually the number of articles refereed in each journal by each referee and including this information as a matter of course as part of an individual's professional credentials.)
5. The job of classifying articles shall be given to the most junior member of any faculty, or preferably an unfortunate librarian or secretary who has no particular experience in the subject area, except that they have been informed of what is the current collection of O.K. words for capitalization.

Footnotes are part of an associative system for providing evidence for research. A plethora of footnotes sometimes may well-indicate that the individual is enamored with secondary sources or that he feels that a listing of great names will help support an argument that is lacking either in theoretical structure or in first-hand empirical evidence.

Costs have been discussed in Chapter 4 but at this point it is worth stressing that the literature search can be costly in personal time, and the structure of rewards is designed to give very little to those individuals who spend much of their time providing services for others. A young scientist or scholar may not find it profitable to be too helpful to others unless his ambition is to eventually become a dean.

The sociological conditions for scientific research vary from time to time and community to community. During some periods the scientific tribes may pay high tribute to the totem of "objectivity." The price paid for pursuit of *reinewissenschaft*, or pure knowledge is the price that was paid by many Central European universities where until a student had proceeded sufficiently high up the academic pecking order he would have been foolhardy to even attempt to give a qualified opinion on anything.

Especially in the development of newly emerging disciplines, it is of extreme importance that individuals be willing to make evaluations of the quality of the literature. This is not an appeal for irresponsibility. On the contrary, it is the reverse. When one suggests that a piece of literature is good or bad, reason must be supplied to back up the assertion. In a scientific community as well as in any other community, it is only a fool who goes out of his way to make unnecessary enemies. Unfortunately, it is not possible to be all things to all persons all of the time. Honest disagreements in the evaluation of the quality of work and the type of problems selected in any human endeavor need to be encouraged.

Professionals have egos which are either like the egos of anyone else, or maybe bigger. No one in particular likes to read a review which states that his or her work is bad. Nevertheless, it is desirable to try to bring about an atmosphere that makes this possible and minimizes the personal aspects to the criticism.

Anonymous refereeing and reviewing have been used heavily in order to encourage individuals to say what they believe without incurring too much direct personal hostility. Unfortunately, there are two other aspects to anonymous refereeing and reviewing which go in a less useful direction. Because the refereeing or reviewing may be anonymous, unless the individual has a highly personal style, he will truly be the unknown soldier and, especially if he is young, the value of this occupation to his career may not be terribly high. This, of course, is not totally true. For a young man to be selected as a referee by the editor of a prestigious journal is undoubtedly an academic or professional honor that carries some recognition with it. His optimal strategy is to be extremely conscientious and good on the refereeing that he does, but to make sure that he is slow enough so that he does not get sent too many papers. The second dysfunctional feature to anonymous refereeing, reviewing, or categorizing is that the anonymity provides a cover for inefficiency and arbitrary action. If the individual knows that he is going to put out a product without his name attached to it, he need not take the responsibility for its quality in the way that he would have to if his name were attached to the work.

A CLASSIFICATION SCHEME FOR
A BIBLIOGRAPHY ON GAMING

In previous joint work with G. Brewer,[1] a classification scheme for articles on gaming and some allied topics was constructed to be used to build up a computerized reference system. A modified and shortened form of this classification scheme is indicated in Table 7.1.

A game's primary purpose is not always evident, nor is the stated purpose necessarily the real one; frequently games have more than one purpose. Hence, the categories offered here are not meant to be mutually exclusive.

The category, "article type," includes as 3-2 "allied topic, relevant." There are many articles involving simulation and model-building that are relevant to gaming but cannot be strictly categorized as such. It is desirable to leave a certain amount of leeway in a bibliography to connect into the more important examples of allied literature.

"Mathematical sophistication" is a category that may be of some importance to a large class of readers. It is extremely frustrating to

TABLE 7.1   A Classified and Evaluated Bibliography on Gaming

Descriptor

ENTRY HEADINGS		ENTRY HEADINGS	
1-0	Purpose: Other	2-7	Subject: Economics
1-1	Purpose: Teaching	2-8	Subject: Education
1-2	Purpose: Training	2-9	Subject: Bargaining/ Bidding
1-3	Purpose: Operational	2-10	Subject: Game Theory
1-4	Purpose: Experimental	2-11	Subject: Gaming Theory, Methodology & Technology
1-5	Purpose: Entertainment		
1-6	Purpose: Research/Theory Development		
1-7	Purpose: Popularization, Advocacy	2-12	Subject: Urban Planning/ Ecology
		2-13	Subject: War Gaming
2-0	Subject: Other		
2-1	Subject: Business/or		
2-2	Subject: Political-Diplo-matic-Military	3-1	Article Type: Discursive
2-3	Subject: Social Psychology/ Psychiatry	3-2	Article Type: Allied Topic, Relevant
2-4	Subject: Artificial Intelligence	3-3	Article Type: Game Description
2-5	Subject: Political Science	3-4	Article Type: Game Documentation
2-6	Subject: Sociology/Organization Theory	3-5	Article Type: Game Results

TABLE 7.1   A classified and Evaluated Bibliography on Gaming—Continued

Descriptor

ENTRY HEADINGS		ENTRY HEADINGS	
4-1	Mathematical Sophistication: 1. None	7-4	Qualitative Assessment: 5. Limited Use 7-8
4-2	Mathematical Sophistication: 2. Slight	7-5	Qualitative Assessment: 6. Bad 9-10
4-3	Mathematical Sophistication: 3. Moderate	10-0	Funding Source: NA
4-4	Mathematical Sophistication: 4. Middling	10-1	Funding Source: ARPA
		10-2	Funding Source: JCS
4-5	Mathematical Sophistication: 5. High	10-3	Funding Source: USA
		10-4	Funding Source: USAF
		10-5	Funding Source: USN
5-1	Type Publication: Book	10-6	Funding Source: Other DOD
5-2	Type Publication: Bibliography	10-7	Funding Source: Other US Government
5-3	Type Publication: Journal Article/Chapter in a Book	10-8	Funding Source: University
		10-9	Funding Source: Foundation
5-4	Type Publication: Technical Report	10-10	Funding Source: Private (Business, Self, Misc.)
5-5	Type Publication: Unpublished Paper	10-11	Funding Source: Other
5-6	Type Publication: Popular Article/Magazine	11-1	Year Published: 1945 or Before
		11-2	Year Published: 1946-1950
7-0	Qualitative Assessment: 1. Not Evaluated 0	11-3	Year Published: 1951-1955
		11-4	Year Published: 1956-1960
7-1	Qualitative Assessment: 2. Excellent 1-2	11-5	Year Published: 1961-1965
7-2	Qualitative Assessment: 3. Of Note 3-4	11-6	Year Published: 1966-1970
		11-7	Year Published: 1971-
7-3	Qualitative Assessment: 4. Modal 5-6	12-1	Game Name ( )
		13-1	Game Facility ( )

Note: The numbers to the left one for computer coding. The numbers to the right of the entry headings 7 = 0–7 = 5 are for evaluation, and the abbreviations in the entry headings No's. 10 are for various government agencies. N.A. means "not available." Categories 6, 8 and 9 have been deleted.

be confronted with an article that has a popular title such as "How to Cut a Sandwich," or "The Mating Problem," and to find that it is an article in topology or combinatorics that requires a PhD in mathematics.

The category "qualitative assessment" was included in order to provide a subjective guide to the articles we have read. Hopefully, at some point it would be possible to have a set of articles evaluated by a large enough group of peers to see what sort of frequency emerges from the qualitative assessments of a group of individuals reading the same literature. A single numerical scale is not a sophisticated assessment scheme, and should not be interpreted as such. The five-point (or ten-point) scale is merely provided as a crude assessment of quality.

The funding source for the work is given when available. In the sociology of science, much can be learned from noting where the resources are being allocated. This relatively elaborate scheme is not fully utilized here, but it is noted to indicate the type of bibliographic and evaluative work that is being done.

SOURCES FOR GAMING LITERATURE AND INFORMATION

Where should the would-be user of gaming look for sources of information and for games, models, or simulations to use? There are several basic sources: certain journals, certain institutions, bibliographies that are either separate publications or are part of existing books, and a number of reference books. The reference books will be noted again in subsequent chapters dealing with specific types of gaming. The other sources are dealt with only here.

Journals

The only journal devoted almost exclusively to gaming is *Simulation and Games*, published by Sage Publications. It deals primarily with games for teaching and includes both theoretical and experimental results. A valuable section having both book reviews and reviews of actual games is included. There is a section announcing the availability of new games and simulations. In 1973, it had a circulation of around 1500 copies.

For those interested primarily in experimental gaming with an emphasis on game theory rather than sociophychological properties of the players, the more important journals are: *Conflict Resolution*, *Management Science*, and the *International Journal of Game Theory*. Possibly the largest and most consistent source of reports on experimental games is *Conflict Resolution*, which until recently had a special section entitled "The Gaming Section," and even now still reports two or three articles per issue on experimental games. The articles are directed toward political science, international problems, and conflict resolution in general. Nevertheless, there is a stress on the relationship between some of the games and the suggested game-theoretic solution for them. Articles on experimental games more closely related to game theory occasionally appear in the *International Journal of Game Theory* and in *Management Science*. In the later, stress is more toward business games and bidding. Other sources for the occasional article on gaming related to economics, bidding, or allied topics are *Decision Sciences*, the *Quarterly Journal of Economics*, the *Journal of the Operations Research Society of America*, and the *Naval Logistics Research Quarterly*.

Further sources for reports on political gaming and on the more hard-core military gaming include: *World Politics*, and the *American Political Science Review*, for the former; and the *Journal of the Operations Research Society of America*, the *Naval Logistics Research Quarterly*, the *Air University Review*, the *Canadian Army Journal*, and many other military journals, for the latter. There is also a literature for the amateur war gamer (estimates higher than mine suggest that these are up to 40,000 in the United States). This includes the publications, *The General* and *Tactics*.

There also exist less-formal publications, such as *Simulation*, the technical journal of the Simulation Councils, and occasional memoranda from the College or Gaming and Simulation of the Institute of Management Sciences.

For those interested in gaming with an emphasis on social psychology, there are many journals which publish this literature. They include: the *Journal of Psychology*, the *Journal of Experimental Psychology*, *Behavioral Science*, the *Journal of Experimental and Social Psychology*, the *Journal of Abnormal and Social Psychology*, the *Journal of Personality and Social Psychology*, the *Administrative Science Quarterly*, and the *American Behavioral Scientist*.

Institutions and Associations

There are only a handful of institutions or organizations actively involved in gaming. The major private, or nonprofit firms include: Rand, Research Analysis Corporation (RAC), Stanford Research Institute, and the Institute for Defense Analysis.

Among the associations partially or completely concerned with gaming are: the Military, Operations Research Society (MORS), the Gaming and Simulation College of the Institute for Management

Sciences, and the National Gaming Council. There are also several small firms engaged in various aspects of manufacturing or promoting games primarily for educational purposes, or as aids to social problem solving. They are noted here not as an endorsement of quality, but because they have existed long enough to be considered as sources of interest. They include: Abt Associates; Academic Games Associates, a group interested in developing and evaluating games for basic education; the games group at the Mental Health Research Institute, University of Michigan, led by Layman E. Allen and devoted to development and testing of instructional games in mathematics, logic, and other subjects. Learning Games Associates is the small firm which has produced many of the games developed by Layman Allen.

The Avalon Hill Company (which publishes *The General*) is a well-known commercial firm that manufactures high-quality war games played by many of the amateur war gamers. There are several amateur war-gaming associations, such as the International Federation of War Gaming.

Among the more important military organizations involved in gaming are: the Joint Chiefs of Staff Studies Analysis and Gaming Agency (SAGA); the Army's Strategy and Tactics Analysis Group (STAG); the Naval War College; and, for work on human factors analysis relevant to gaming and simulation, the Behavioral Science Research Laboratory (BSRL) of the Army; and a group in Naval Personnel.

Among the sources for non-United States activities are: the operational research and gaming group of the Admiralty at Byefleet in England; and the Rand Corporation's occasional surveys of Soviet literature.

Bibliographies and Books

There are a large number of bibliographies in various subjects associated with gaming and simulation. Many are of relatively low quality in that they appear to be little more than a hastily compiled cut-and-paste job with little classification and frequent errors, including items classified by title only, where a perusal of the actual paper would have indicated that it had very little to do with the actual title. A listing and discussion of a few of the better and/or better-known bibliographies is given.

The following categorizations are used, where possible, in the description of the bibliographies:

A = Annotated
E = Evaluated
C = Categorized

By "categorized" I mean that the description of the article is broken down to include more specification than is supplied merely by the title or the major subject divisions of the bibliography.

In the following chapters references are given for suggested reading. Full references to the books noted in this chapter are given in the subsequent listing of books on the specific topic. However, these are not intended to be anywhere near as comprehensive as many of the bibliographies noted below.

Educational Gaming

Twelker, P. A., *Instructional Simulation Systems:* An Annotated Bibliography Corvallis, Ore.: Continuing Education Publications, 1969 (A, C). This is a guide for the potential game user to 404 games with a supplementary list of about 450 other games either discontinued or in development at the time of publication. It provides a brief description and references to availability, information on playing time, number of players, preparation, materials needed, and costs for all of the games described. These are combined with some fairly cryptic but useful comments on the type of game, occasionally with a hint of the reviewer's evaluation of the game. This publication would have been even more valuable than it is had there been more in the way of evaluative remarks. All of the games covered are for educational purposes. The topics range from general business, specific industry models, functional models, games directed toward political processes, ecology, international relations, urban affairs, sociology, psychology, and skill-development games. There is no theoretical discussion accompanying the listing. The book is a "where to find it" and "what it is" publication.

Kidder, S. J., "Simulation Games: Practical References, Potential Use, A Selected Bibliography," *Report No. 112*, Baltimore: Center for Social Organization of Schools, Johns Hopkins, August 1971. This examines various books, research studies, and demonstration projects concerning the potential of simulation and gaming for teaching and training and for the investigation of social and psychological processes. It contains a bibliography of 113 publications on gaming for educational purposes. There is a brief essay on the potential of gaming, and the references on gaming for educational purposes are reasonably carefully selected.

Gibbs, G. I. (Ed.), *The Handbook of Games and Simulation Exercises*, Beverly Hills, Calif.: Sage, 1974; contains a good Coverage of gaming activity in Great Britain, among its other features. In this publication around 1,500 games are noted. The addresses of various amateur gaming groups and societies as well as commercial enterprises are supplied. This is a reasonably useful service publication, but it is eclectic in style and offers no quality guide. The references included range from extremely good to extremely poor, and there is no indication given to the reader as to how to distinguish among them.

Another source for individuals who wish to select for education purposes is the book edited by D. Zuckerman and L. Horn entitled *A Guide to Simulation Games for Education and Training* Cambridge Mass.: Information Resources, Inc., 1970. (A, C). It describes in reasonable detail around 400 educational games.

The *American Behavioral Scientist 10*, 3 (November 1966) published a selective bibliography on simulation games as learning devices (pp. 34—36). Although many good references are included, this is a relatively small, eclectic set of references which would be of little use to someone who does not already have some background and sense of discrimination in his reading on the uses of games for teaching.

Business Games

Business gaming has become a sufficiently accepted activity over the last 15 years that there is a considerable literature with many books and references. Virtually any of the books devoted to the subject contains bibliographies of one form or the other. In particular, the book by Graham, R. G., and Gray, C. F., *Business Games Handbook* (A, C), put out by the American Management Association in New York in 1969, contains descriptions of around 200 business games. The book is expensive ($22.00) and, although it has some introductory reading and discussion of the general purposes and difficulties of business games, it is primarily a "where to get" and a "how to" reference book. It contains a well-chosen bibliography with 295 references, primarily to business games.

The words "games" and "simulation" are frequently used interchangeably. Hence, the reader must be forewarned that the title containing the word "simulation" may be either about strictly com-

128

puter planning models with no human participation or about gaming exercises. For example, *Business Simulation Industrial and University Education*, Englewood Cliffs, N.J.: edited Prentice Hall, 1962, by Greenlaw, H., Herron, L. W., and Rawdon, R. H. begins with the statement, "We define a business simulation as a sequential decision-making exercise structured around a model of a business operation in which participants assume the role of managing the simulated operation. We use the term, 'simulation,' and 'game' interchangeably." This is a book dealing primarily with gaming. On the other hand, *Simulation in Business and Economics*, by Meier, R. C., Newell, W. T., and Pazer, H. L., Englewood Cliffs, N.J. Prentice-Hall, 1969 has one chapter out of nine dealing with gaming. This latter book contains many useful references to simulations and models of business and economic processes and institutions. For those interested in seeing the type of corporate simulations that have been built, the book edited by A. N. Schrieber entitled *Corporate Simulation Models*, Seattle: University of Washington, 1970, even though it lacks an index and is the result of a conference, contains a broad array of the type of models being built. Other bibliographies focusing primarily on business or management simulations include Malcolm, D. G., "A Bibliography on the Use of Simulation in Management Analysis," *Operations Research*, 8, 2, (March-April, 1960), and a more-recent "Bibliography 19: Simulation and Gaming," by Naylor, T. H., *Computing Reviews*, (January 1969), 61-69. Both of these range further afield than the subjects of business and economics. Unfortunately, neither offers further classification, nor evaluation.

Experimental Games

"Experimental Games: A Bibliography (1945-1971)," Ann Arbor: Mental Health Research Institute, The University of Michigan Communication #293, March 1972 put together by M. Guyer and B. Perkel, contains around 1,000 references, most of which are directed to reports on experiments. This is an extremely useful bibliography for the researcher. However, it also contains a fair number of references to articles in game theory and a few references to other topics. The bibliography would be considerably improved if these had been separated out and classified in a different manner. The authors admit that their labor of love was done with the motivation that can be best summed up in their own words, "It is hoped that the compilation of this bibliography will serve to establish a base upon which to build an information storage and retrieval system for published papers in the area of experimental gaming." (p. 5)

Limiting our interest to experiences with 2 × 2 matrix games, the forthcoming book by Rapoport, A., Guyer, M. J., and Gordon, D. J., *The 2 × 2 Game*, contains the most exhaustive set of references to experiments with 2 × 2 matrices that exist.

Naylor, T. H., "Experimental Economics Revisited," *Journal of Political Economy*, *80*, 2, (March-April 1972), 347–352, contains around 30 references to experimental gaming in economics.

There is also a useful but unpublished bibliography on experimental games involving coalition formation compiled by J. W. F. Throop, Department of Psychology, Carleton University.

The book by Shubik, M., *Games for Society, Business and War*, Amsterdam: Elsevier, 1975, also contains many references to the experimental literature.

Game Theory

There are now thousands of references to the theory of games, most of them not of direct interest to those involved in gaming. However, a certain familiarity with the concepts of game theory is required for much of experimental gaming and certainly some military and political science gaming. An adequate set of references to the essential ideas in game theory for the individual interested in gaming are to be found in the following four books: Shubik, M. (Ed.), *Game Theory and Related Approach to Social Behavior*, New York: Wiley, 1964. This book contains an introductory essay and an annotated bibliography (A, C); Rapoport, A., *Two Person Game Theory: The Essential Idea*, Ann Arbor: University of Michigan Press, 1964; Davis, M. D., *Game Theory—A Nontechnical Introduction*, New York: Basic Books, 1970; Shubik, M., *Games for Society, Business and War*, Amsterdam: Elsevier, 1975. The game theory coverage in this last book is aimed specifically at those interested in gaming.

Military Gaming

Although it is somewhat out of date, Riley, V., and Young, J. P., "Bibliography on War Gaming," Baltimore: Operations Research Office, Johns Hopkins University, (April 1, 1957) (A, C), is still a useful reference which is partially annotated and broken into subclassifications.

A somewhat more up-to-date source is provided by F. J., McHugh, *Fundamentals of War Gaming*, Newport, R.I.: U.S. Naval War College (3rd. ed.), March 1966.

For those interested in an overall survey of the uses and purposes of military models, simulations, and games, the study by Martin Shubik, M., and Brewer, G. D., "Models, Simulations, and Games," Santa Monica: Rand Corporation, R-1060-ARPA/RC, May 1972 (A, E, C), contains a survey of 132 different models, simulations, or games.

Another source which, unfortunately, for obvious reasons is for restricted or official use only is the *Catalogue of War Gaming Models*, occasionally published by the Joint War Games Agency for the Joint Chiefs of Staff in Washington. This publication contains, in general, less information than an operational gamer would like, but more information than is usually available to those outside the profession.

For those interested in war games, it should be noted that most Department of Defense operational war games fall into different levels of classification. In general, an overall description of the game and its purposes is usually not classified and can be found by an outsider by sufficient digging. In some instances, certain parts of the actual structure of the model will be classified as restricted or for official use only. Almost always any specific numbers, parameters, or estimates will be highly classified and will not be available to those without clearance.

There is a specialized literature on two-person game-theory relevant to military gaming. This involves the study of duels, games of pursuit, and interception. Many references to this literature are to be found in Dresher, M., *Games of Strategy*, Englewood Cliffs, N.J.: Prentice Hall, 1961 (also available in Russian).

A popular and general set of references to military gaming of any variety can also be found in the book by Wilson, A., *The Bomb and the Computer*, New York: Delta, 1968.

Political Science and International Relations

Since the inception of the Inter-Nation Simulation, first constructed by H. Guetzkow, his students, and his colleagues in 1957, publications on gaming in political science and international relations have mushroomed.

A rich but by now somewhat outdated source of references is the book by Guetzkow, H., Alger, C. F., Brody, R. A., Noel, R. C., and Snyder, R. C., *Simulation in International Relations*, Englewood Cliffs, N.J.: Prentice Hall, 1963. Some more recent references can be found in Laponce, J. A., and Smoker, P., *Experimentation and Simulation in Political Science*, Toronto: University of Toronto, 1972.

An unpublished bibliography prepared by E. Mickolus at Yale University Department of Political Science in 1974 contains around 750 references, most of them pertaining directly to political science and international relations.

Simulation

As has been noted in Chapter 1, a knowledge of some of the aspects of game theory, model building, and simulation is highly desirable, if not vital, for those who wish to work with games for any purpose. Model building and simulation are probably still more an art than a science. Nevertheless, it is desirable for those who have to construct or understand the simulated environments in which games are played to appreciate the technique involved.

There are several broad, general bibliographies on simulation. One of the earliest with several-hundred references is Shubik, M., "Bibliography on Simulation, Gaming, Artificial Intelligence," *Journal of the American Statistical Association*, 736-751, (December 1960), *55* This is clearly somewhat out-of-date; and Naylor, T.H., "Simulation and Gaming" *Computing Reviews, 10,* 1 (January 1969), 61–69, contains many more recent references, complete with a breakdown of topic headings into general textbooks, methodology, simulation languages, experimental design, data analysis, etc.

Another extremely useful source, especially for those who do not wish to enter too deeply into the problems of simulation, and yet want a reasonable coverage are the references supplied by Barton, R. F., *A Primer of Simulation and Gaming*, Englewood Cliffs, N.J.: Prentice Hall, 1970.

General References on Gaming and Allied Topics

Shubik, M., Brewer, G., and Savage, E., *The Literature of Gaming, Simulation, and Model Building: Index and Critical Abstracts*, Santa Monica: Rand Corporation, R-620-ARPA (June 1972) (A, E, C), provides a categorized, partially abstracted, and evaluated reference to around 1,000 articles, books, and games. The associated publication, Shubik, M., and Brewer, G., *Reviews of Selected Books and Articles on Gaming and Simulation*, Santa Monica: Rand Corpora-

132

tion, R-732-ARPA (June 1972), contains reviews of around 50 books related to gaming.

There are other subjects that the dedicated gamer may find himself interested in: for example, the relationship between gaming and sports, or the use of games in psychiatry, diagnostic treatment, and care programs. Excellent bibliographies on both of these topics may be found in Avedon, E. M., and Sutton-Smith, B., *Study of Games*, New York: Wiley, 1971.

Those interested in board games will find a lengthy listing and an excellent set of references in the scholarly work of Murray, H. J. R., *A History of Board Games Other Than Chess*, London: Oxford University Press, 1952. For the buff who truly enjoys obscure information, in Bell, R. C., *Board and Table Games*, London: Oxford University Press, 1960, a biography of ten of the more important students of games is supplied, starting with As-Suli who wrote the first known book on chess sometime around 920A.D.

Human factors analysis and the possibility of the construction of robots are both of more than passing interest to those concerned with both experimental and operational games. There is a vast literature on human factors. In case the reader is numbed by now, only one bibliography is referred to: Air University, School of Aviation Medicine, USAF "Fact and Fancy in Sensory Deprivation Studies," *Aeromedical Reviews, Review 5–59*, Austin, Tex., Brooks Air Force Base, (August 1959). It contains 211 references on human performance, fatigue, effect of drugs on fatigue, starvation, volunteering, survival experience, effects of isolation, and so forth.

Unfortunately, the direction of current work has emphasized the dichotomy between the computer scientist's approach to artificial intelligence and the behavioral scientist's approach to the construction of artificial players and stooges in gaming experiments. Even though they are both somewhat out of date, the reader's attention is called to the excellent bibliography by Minsky, M., "A Selector Descriptor-Indexed Bibliography to the Literature on Artificial Intelligence" *IRE Transactions on Human Factors in Electronics*, (March 1961), 39–55; and to Feigenbaum, E. A., and Feldman, J., (Eds.) *Computers and Thought*, New York: McGraw-Hill, 1963. This also contains an excellent bibliography.

REFERENCE

1. Shubik, M., Brewer, G., and Savage, E., The Literature of Gaming, Simulation and Model Building: Index and Critical Abstracts, Santa Monica: Rand Corporation, R-620-ARPA (June 1972).

8 Gaming for Business, Management Operations Research, and Economics: A Literature Guide

In this and each of the subsequent brief chapters, the following outline is adopted:

1. purpose and type
2. usage
3. a bibliography
4. commentary
5. examples

Under the first title a statement of the purposes for gaming is given. The dysfunctional reasons already discussed in Chapter 6 will not be referred to again here. Also discussed in this section is a breakdown of games into the several types that may be relevant to this type of gaming.

"Usage" includes an indication of what sort of individuals are using this type of gaming and whether or not the use is widespread. The bibliography that follows contains references from two to ten books and five to twenty articles. The commentary section that follows discusses many of these books and articles. When relevant and feasible, an example of a game falling in the category being discussed is given; and information is supplied on course outlines or the organization of teaching.

PURPOSE AND TYPE

The major uses of gaming in business management and operations research have been for teaching and training. There has been little gaming for serious operational purposes and although there has been some experimental gaming, the relative emphasis is such that a discussion of experimental gaming is relegated to the next chapter. The above remarks hold almost as well for gaming activities related directly to economics, with the possible difference being that there has been a somewhat higher interest in experimental gaming in economics.

The major categorization in terms of type of game has been into (1) general purpose games with no direct empirical industry association; (2) special purpose and functional games designed to teach individuals or train them in certain specific skills, such as accounting, inventory management, logistics, or personnel relations, etc.; and (3) games specifically associated with an industry. Thus, for example, there are games concentrating on aerospace, banking, insurance, retailing, transportation, and many other industries.

Usage

The major users of business management and O. R. gaming are colleges and universities at both the undergraduate and graduate levels and corporate and military training programs. There are several games designed for high-school and grade-school usage, but these form a relatively small part of the games being utilized. A survey by J. D. Couger of 139 member schools of the American Association of Collegiate Schools of Business indicated that the general business game is now part of the business policy course in 45 percent of the schools. The major use is at the undergraduate level where, for example, in marketing 38 percent of the schools use gaming at the undergraduate level, 26 percent at the master's level, and only 4 percent at the PhD level. In finance, the percentages are 22 percent, 15 percent, and 2 percent; and in management, 38 percent, 31 percent, and 7 percent. Couger noted that between 1966 and 1972 there has been not only considerable growth in the use of gaming among business schools, but there has also been a change in the approach. In particular, the use of gaming has become far more integrated into teaching in various courses than had originally been the case.

A BIBLIOGRAPHY

Books

Gaming

Barton, R. E., *A Primer on Simulation and Gaming*, Englewood Cliffs, N.J.: Prentice Hall, 1970.

Carlson, J. G. H., and Misshauk, M. J., *Introduction to Gaming: Management Decision Simulations*, New York: Wiley, 1972.

Churchill, N. C., Miller, M. V., and Trueblood, R. M., *Auditing Management Games and Accounting Education*, Vol. II, Contributions to Management Education Series, Homewood, Ill.: R. D. Irwin, 1964.

Cohen, K. J., Dill, W. R., Kuehn, A., and Winters, P. R., *The Carnegie Tech Management Game: An Experiment in Business Education*, Vol. I, Contributions to Management Education Series, Homewood, Ill.: R. D. Irwin, 1964.

Graham, R. G., and Gray, C. F., *Business Games Hand Book*, New York: American Management Association, 1969.

Greene, J. R., and Sisson, R. L., *Dynamic Management Decision Games*, New York: Wiley, 1959.

Greenlaw, P. S., Herron, L. W., and Rawdon, R. H., *Business Simulation in Industrial and University Education*, Englewood Cliffs, N.J.: Prentice Hall, 1962.

Haldi, J., and Wagner, H. M., *Simulated Economic Models*, Homewood, Ill.: R. D. Irwin, 1963.

Henshaw, R. C., and Jackson, J. R., *The Executive Game*, Homewood, Ill.: R. D. Irwin, 1966.

Kibbee, J. M., Craft, C. J., and Nanus, B., *Management Games*, New York: Van Nostrand Reinhold, 1961.

Levitan, R. E., and Shubik, M., *The Financial, Allocation and Marketing Executive Game*, Sands Point, N.Y.: IBM Executive Department, 1961.

Lindert, P. D., *MACRO: A Game of Growth and Policy*, New York: Holt, Rinehart, and Winston, 1970.

McFarlan, F. W., McKenney, J. L., and Seiler, J. A., *The Management Game: Simulated Decision Making*, New York: Macmillan, 1970.

McKenney, J. L., *Simulation Gaming for Management Development*, Cambridge: Harvard Business School, 1967.

Thorelli, H. B., Graves, R. L., and Howells, L. T., *International Operations Simulation with Comments on Design and Use of Management Games*, New York: Free Press, 1964.

Simulation

Amstutz, A. E., *Computer Simulation of Competition and Market Research*, Cambridge, Mass.: M.I.T. Press, 1967.

Balderston, F. E., and Hoggatt, A. C., *Simulation of Market Processes*, Berkeley: Iber Special Publications, University of California, 1962.

Bonini, C. P., *Simulation of Information and Decision Systems in the Firm*, Englewood Cliffs, N.J.: Prentice Hall, 1963.

Clarkson, G. P. E., *Portfolio Selection, A Simulation of Trust Investment*, Englewood Cliffs, N.J.: Prentice Hall, 1962.

Forrester, J. W., *Industrial Dynamics*, New York: Wiley, and Cambridge, Mass.: M.I.T. Press, 1961.

136

Naylor, T. H., Balinfy, J. L., Burdick, D. S., and Chu, K., *Computer Simulation Techniques*, New York: Wiley, 1966.

Orcutt, G. H., Greenberger, M., Korbel, J., and Rivlin, A. H., *Microanalysis of Socioeconomic Systems—A Simulation Study*, New York: Harper & Row, 1961.

Schrieber, A. M. (Ed.), *Corporate Simulation Models*, Seattle: University of Washington Printing Plant, 1970.

Articles

Gaming

Bastable, C. W., "Business Games, Models, and Accounting," *Journal of Accounting, 109* (March 1960, 59–60.

Campbell, F. M., and Ashworth, E. R., "Monopologs: Management Decision Making Game Applied to Tool Room Management," *The Journal of Industrial Engineering, XI*, 5 (September-October 1960), 372–377.

Cohen, I. K., and Van Horn, R. L., "A Laboratory Exercise for Information System Evaluation," in Spiegel, J., and Walker, D. (Eds.), *Information System Sciences*, Washington, D.C.: Spartan Books, 1965, pp. 321–341.

Cohen, K. J., and Rhenman, E., "The Role of Management Games in Education and Research," *Management Science, 7* (1961), 131–166.

Deep, S. D., Bass, B. M., and Vaughan, J. A., "Some Effects on Business Gaming of Previous Quasi T-Group Affiliations," *Journal of Applied Psychology, 51*, 5 (1967), 426–431.

Dennick, W. H., and Olanie, F. R., "Bank Management Game," *Bankers Monthly, 77* (September 1960), 56–60.

Dill, W. R., "What Management Games Do Best," *Business Horizons, 4* (Fall 1961), 55–64.

Dolbear, F. T., Attiyeh, R. R., and Brainard, W. C., "A Simulation Policy Game for Teaching Macroeconomics," *Papers and Proceedings of the 8th Annual American Economic Association Meetings*, Washington, D.C. (December 1967), 458.

Eiolart, T., and Searle, N., "Business Games Off the Shelf," *Simulation, 20*, 2 (February 1973), 63–66.

Feeney, G. J., "Simulating Marketing Strategy Problems," *Marketing Times, 2*, 1, January 1959, 8–24.

Geisler, M. A., "The Simulation of a Large-Scale Military Activity," *Management Science, 5*, 4 (July 1959), 359–368.

Kennedy, M., "A Business Game for Accountants," *Journal of Accounting, 73*, 823 (March 1962), 219–222.

Levitan, R. E., and Shubik, M., "A Business Game for Teaching and Research Purposes," Yorktown Heights, N.Y.: IBM, Thomas J. Watson Research Center (1962), RC 730 and RC 731.

McKenney, J. L., "An Evaluation of a Business Game in an MBA Curriculum," *Journal of Business, 35* (July 1962), 278–286.

McRaith, J. F., and Goeldner, C. R., "A Survey of Marketing Games," *Journal of Marketing, 26,* 3 (July 1962), 69–72.

Mellor, P., and Tocher, L., "A Steelworks Production Game," *Operational Research Quarterly, 14,* 2 (June 1963), 131–136.

Roberts, A. L., "What's Wrong with Business Games?" *Journal of Industrial Engineering, 13,* 6 (November-December 1962), 465.

Shubik, M., Wolf, G., and Lockhart, S., "An Artificial Player for a Business Market Game," *Simulation and Games, 2,* 1 (March 1971), 27–43.

Watkins, H. R., "Business Games in Business," *Operational Research Quarterly, 10,* 4 (1959), 228–244.

Wikstrom, W. S., "The Serious Business of Business Games," *Management Record* (February 1960), 6–25.

COMMENTARY

Books

Gaming

Barton, *A Primer on Simulation and Gaming*: This is a useful, simple introductory text that stresses simulation more than it does gaming. This short primer is simply written, and makes no attempt at philosophy. Simple exercises are appended to each chapter and reflect the content of the chapter. The book provides a discussion of simulation models and theory, indicates the use of simulation as a research tool, and notes the various techniques of simulation. A chapter is devoted to computer systems, including input, output media, memories, and the system as a whole. Man computer simulation and all computer simulations are discussed in two chapters. There is a chapter on Monte Carlo techniques followed by a chapter on simulation languages. The last chapter, which provides a survey of simulation applications, is relatively weak.

Cohen et al., *The Carnegie Tech Management Game*: This describes one of the more important larger, and long-used business games. It is important reading for those who wish to integrate a large-scale business game into a graduate program using games in conjunction with many different teaching applications. Games of the size and complexity of the Carnegie Tech Game are more than mere exercises. They become virtually institutions in their own right. In considering games of this size, it is best to think in terms of a teaching and experimental program rather than in terms of a game. The mere availability of the computer program is only a small fraction of that which is required for successful application.

138

Churchill et al., *Auditing Management Games and Accounting Education:* This is a companion to *The Carnegie Tech Management Game* and describes experiences in using the game as a context for accounting education. In particular, the role of the audit is discussed. The general conclusions of the authors were that the business game provided a highly successful supplementary teaching device for use in accounting education.

Graham and Gray, *Business Games:* This is a relatively expensive, reasonably well-done "how-to and where-to-find" survey. It is a fairly straightforward, competent production.

Haldi and Wagner, *Simulated Economic Models:* This book attempts to apply gaming to teaching of microeconomic theory. It is geared to liberal arts, business, and engineering students and presents a series of six laboratory exercises designed to teach certain economic principles. It does not require a computer. The exercises are designed to stress the utility of economic theory applied to a variety of complex and reasonably realistic problems. Dynamics and the roles of uncertainty and risk are stressed. This book represents a first-class effort in designing a serious game for teaching. The structure is simple; the game is easy for the administrator to control; and there is rapid feedback in the form of immediate tests on comprehension. Furthermore, the length of the sessions is appropriate. The authors used the game for several years with considerable success before publishing the final version. It seems odd that, given its merit, it is scarcely in use today. Its success in the hands of the authors may have been due to their unique motivation and skill. Its lack of success with many others may point up the importance of devising procedures to motivate and make it easy for others to use well-designed games. Possibly the most important lesson to be learned is that teaching, like many other trades, must depend not merely upon the superior practitioner but the average practitioner. If the new materials offered are not especially convincing or convenient for the average teacher they may either not be used at all or be poorly used.

Henshaw and Jackson, *The Executive Game:* This describes the Executive Game, which is a direct descendant of the UCLA Executive Game No. 2. This latter game is one of the best known and most widely used of business teaching games. The distinction of this book and of the UCLA project in general is its thoroughness of documentation. With this single publication the game can be set up, run, and used for teaching purposes by an individual, with only moderate technical qualifications. In other words, the game is easily transferrable and can be used virtually anywhere. In this publication the game is presented with a player's manual, the underlying mathematical models and flow charts are given, and a reproduction of the computer program is provided. Among the strategic variables included are price, marketing expenditures, R and D, maintenance, volume of production, capital investment, purchase of raw materials, and setting of dividends. The output format is more or less in terms of standard business reporting methods. As is almost always the case, the success in the use of the game depends somewhat upon the game administrator's imagination and skill. The experimental potential for a game such as this is quite high yet it appears that it has scarcely been used for experimental purposes.

Simulation

Balderston and Hoggatt, *Simulation of Market Processes*: This is an all-computer simulation and a theoretical, as well as empirical, study of the process of market sector formation in the lumber industry. It provides an excellent example of a mixture of economic investigation and analysis together with mathematical model-building and simulation experimentation.

Forrester, *Industrial Dynamics*: This is a highly readable and, on first encounter, persuasive plea for broad systems modeling on industrial dynamics systems. The world is neither as simple nor as easy to model as Professor Forrester's book would lead one to believe. Nevertheless, his methods of modeling and the Dynamo language are useful, convenient, and suggestive as a first step in teaching large-scale systems modeling. If used with an understanding of their limitation, it provides a nice introduction to simulation.

Naylor et al., *Computer Simulation Techniques*: This book is accurately described by its title. It is a first-class textbook dealing with the methods and techniques of digital computer simulation.

Orcutt et al., *Microanalysis of Socioeconomic Systems*: This represents a first step in the construction of the type of microeconomic simulation needed to improve economic planning and forecasting. It concentrates primarily upon a demographic model.

Schrieber (Ed.), *Corporate Simulation Models*: Most conference volumes are uneven, and this one is no exception. Nevertheless, it contains several interesting pieces describing models built for planning purposes by many different firms.

None of the books on simulation described above is really of direct primary interest to someone whose concern is gaming. Nevertheless, all are of high secondary importance, as most games are played in simulated environments. A badly simulated environment

140

may totally vitiate any value that a game might otherwise have. Model-building and simulation are still very much of an art, and the gamer must at least be able to know the questions and the problems that confront him in selecting appropriate models or simulations to serve as the basis of the description of the environment for his game.

Articles

Bastable, "Business Games, Models and Accounting": This presents a discussion of the uses of gaming for theory development and training purposes in accounting. The author gives several dramatic illustrations of what different accounting procedures imply for decision making in a business firm. Although its attempt to model some aspects of accounting theory is somewhat crude, the article is important in demonstrating a promising application for games for teaching.

Deep et al., "Some Effects on Business Gaming of Previous Quasi T-Group Affiliations": With the Carnegie Tech Management Game as the experimental vehicle, the authors assigned 93 graduate business students to 9 companies, according to whether they had been in the same or different T-groups in the previous 15-week academic period. The companies composed of players who were separated from their T-group colleagues performed more effectively than those whose T-group associations were left intact. The latter reported less internal conflict but were less effective in playing the game. This paper raises several interesting questions but unfortunately does not provide the answers.

Geisler, "The Simulation of a Large-Scale Military Activity": This article provides a concise description of the mid-project activities at Rand's Logistics Simulation Laboratory. The discussion of Laboratory Problem One notes that the exercise was intended to test alternative logistics policies for the Air Force, to integrate existing research with operational requirements, and to develop man-machine laboratory techniques and facilities. The author notes that the laboratory enterprise raised basic questions about personnel requirements for the conducting of ongoing serious systems simulations. LSL, whose work extended over several years, had 30 professionals and about 20 subordinate staff, a rather elaborate facility, and considerable computational backup.

McKenney, "The Evaluation of a Business Game in an MBA Curriculum": An experiment testing the learning of three concepts was used to compare the performances of MBA students taught by a game and those taught by other methods. The concepts were: (1) today's decisions create tomorrow's environment, (2) goals and plans are carried out by a series of consistent decisions to vary in accordance with the environment, and (3) functional decisions of a firm are interrelated. The tests, statistical procedures, and methods of grading are described in detail. Discussions and conclusions indicate some evidence favoring the use of gaming over more conventional methods. For example, gaming seemed better at showing the interrelationships among functional decisions.

McRaith and Goeldner, "A Survey of Marketing Games": This brief article provides an illustrated list of games available to marketing executives and educators. It is included as an example of a trade publication appearing in a journal where most of the readers have little if any knowledge of the potentials of gaming for their purposes. The article is now clearly out-of-date, and better references are available in the handbooks and bibliographies noted in Chapter 7.

Shubik et al., "An Artificial Player for a Business Market Game": This paper provides a detailed description of the heuristics used to determine the behavior of an artificial player utilized in a business game. Among the advantages in employing an artificial player are control, economy, and the need to explain, justify and contrast the rules of behavior of the artificial player with the behavior of the live players.

EXAMPLES

Many examples can be found described in detail in the books noted above and in Chapter 10 of Shubik: *Games for Society, Business and War*. A brief sketch of one large business game utilized internally by IBM and several of its customers is given below.

The Financial Allocation and Marketing Executive Game (FAME): This game was first written by R. Levitan and M. Shubik at the IBM Research Center, Yorktown Heights, for use in the Executive Development Department Program at IBM, Sands Point. The game was originally designed to be played over a period of one week. It has since undergone several modifications, including the production of a time sharing version, known as the Applied Information and Management Simulation (AIMS). The number of firms is from three to four; the number of products two; the number of market regions three to four; the plant operations are based on two shifts with or without overtime; and the decision period is one quarter. The game was designed to be playable at different levels of complexity in different functions. Thus, for example, in manufacturing, marketing, and finance, decisions could be made at three different decision levels. An example of the buildup in complexity is provided by manufacturing. At the simplest decision level, the player is called up to pick a production level for Product A and Product B. The computer does the rest, purchasing raw materials, assigning labor, and so forth. At the second decision level, the player must make specific decisions concerning the assignment of first and second shifts, the assignment of overtime, hiring, severance, and the ordering of raw materials. At the third level, further decisions are called for. The game has been used to provide a means to demonstrate the value of various management science techniques, including production, inventory scheduling, linear programming, market information processing, and the uses of accounting for control.

Books such as those on the Carnegie Tech Game frequently contain course descriptions. Thorelli and Graves, in the description of INTOP, provide two complete sample course schedules, the first for a course in advanced marketing management, including 22 sessions, and the second for the course in business policy and organization, encompassing 20 sessions. Their stress is on a mixture of lectures, discussions, and the actual playing of the game, together with related assignments. For example, in the Marketing Management Course, 12 quarters of the game are played over the 22 sessions. Lectures between the various quarters include topics such as market channels and systems, sales, forecasting, demand analysis, and the interdependence of market structure and marketing strategy (see pp. 183–191).

The book by Graham and Gray also contains a sketch of the American Management Association's program in gaming.

9 Experimental Gaming: A Literature Guide

One of the fastest-growing areas in the application of gaming has been experimentation. The type of experimentation that has been carried out varies considerably. It can be broken down primarily in terms of major disciplines, with the emphasis being on economics, political science, and social psychology. Another way of considering the breakdown is in terms of games with rich scenarios, games primarily presented in matrix form or in a relatively sparse mathematical form, and games in coalitional form where (in contrast with the matrix game) communication and discussion or other forms of explicit coordination are stressed.

In spite of the considerable use of gaming in the military it is surprising to find so little literature on experimental gaming relevant to military operations research, even though there is a considerable literature on simulation and mathematical models.

USAGE

Most of the users of experimental gaming are at universities or research institutions. The major purposes for their work involve checking behavioral theories and the exploration and generation of new hypotheses concerning behavior. In some cases, this activity may take place simultaneously with other uses of gaming. For example, a game may be simultaneously employed for teaching as well as for experimentation. In other instances, the interest in experimentation may come in trying to establish the validity of the game as either a teaching or an operational device.

In my opinion, it is precisely the need for validation that calls for military gaming experimentation that has not been performed. In the military there appears to be a lack of separate studies about the development of validation criteria. There is a great danger in the assumption that gaming or simulation done for operational purposes will produce its own objective measures of worth.

An indication of the activity in experimental gaming can be obtained by looking at the bibliography compiled by Guyer and Zabner, already referred to in Chapter 7. There are by now literally thousands of experiments with matrix games, and the references in the books listed in the next section attest to the activity in the more specialized disciplines.

The running of experimental games varies in the nature of the facilities and in costs from simple matrix games run in more-or-less borrowed facilities, such as empty classrooms, to highly formalized complex computer programs complete with elaborate laboratory facilities.

It has been noted in Chapter 4 and should be stressed again here that university staffs especially tend to underestimate the cost of their gaming experiments or games used for other purposes. Even a simple matrix game, unless the data are gathered on-line as part of a lecture, tends to cost at least a few hundred dollars if one is paying the subjects, or if one is at least counting the volunteer time as having any value whatsoever. Even elementary data-processing takes several days.

Simple experimental games are nevertheless instructive and fairly easy to run. Hence, they serve as a good teaching device for graduate students. In other words, it is reasonable to encourage graduate students to run simple experimental games both as an educational device for them and as a way for obtaining more experimental evidence. Of course, with this use one trades off certain levels of control for the learning experience obtained by new or tyro experimenters.

A BIBLIOGRAPHY ON
EXPERIMENTAL GAMING

Relative to the number of articles that have been published, the number of books on experimental gaming appears to be somewhat low. However, this may be part of a time-lag phenomenon, and if it is, there should be a considerable increase in the number of books published in the next two or three years. The bibliography presented below groups all of the books together. The articles, however, are classified under five headings: (1) matrix games: two-person, constant-sum games; (2) other matrix games; (3) games with other formats; (4) three-person and larger experimental games; and (5) experimental economics.

Books

Fouraker, L. E., and Siegel, S., *Bargaining Behavior*, New York: McGraw-Hill, 1963.

Rapoport, A., and Chammah, A. M., *Prisoner's Dilemma*, Ann Arbor: University of Michigan Press, 1965.

Rapoport, A., Guyer, M. J., and Gordon, D. J., *The 2 X 2 Game*, Ann Arbor: University of Michigan Press (Forthcoming).

Sauermann, H. (Ed.), *Contributions to Experimental Economics*, Vol. 1, 1967; *Contributions to Experimental Economics*, Vol. 2, 1970; *Contributions to Experimental Economics*, Vol 3, 1972. Tubingen: Mohr.

Schelling, T. C., *The Strategy of Conflict*, Cambridge, Mass.: Harvard University Press, 1960.

Shubik, M. (Ed.), *Game Theory and Related Approaches to Social Behavior*, New York: Wiley, 1964.

Shubik, M. *Games for Society, Business and War*, Amsterdam: Elsevier 1975.

Siegel, S., and Fouraker, L.E., *Bargaining and Group Decision Making: Experiments in Bilateral Monopoly*, New York: Macmillan, 1960.

Tedeschi, J. T., Schlenker, B. R. and Bonoma, T. V., *Conflict, Power and Games*, Chicago: Aldine Publishing Co., 1973.

Articles

Matrix Games: Two Person Constant Sum Games

Brayer, A. R., "An Experimental Analysis of Some Variables of Minimax Theory," *Behavioral Science*, (1964), 9, 33–44.

Burgin, G. H., "On Playing Two-Person Zero-Sum Against Non-Minimax Players," *IEEE Transactions on Systems, Science and Cybernetics*, (1969), SSC-5, 369–370.

Fox, J., "The Learning of Strategies in a Simple Two-Person Zero-Sum Game Without Saddlepoint," *Behavioral Science*, 17, (1972), 300–308.

Frenkel, O., reported in A. Rapoport, M. Guyer, and D. J. Gordon, *The 2 X 2 Game*, Ann Arbor: University of Michigan Press (Forthcoming).

Lieberman, B., "Human Behavior in a Strictly Determined 3 X 3 Matrix Game," *Behavioral Science*, 5, (1960) 317–322.

Littig, L. W., "Behavior in Certain Zero Sum Two-Person Games," *Journal of Game*," *Behavioral Science*, 5, (1960) 317–322.

Morin, R. E., "Strategies in Games with Saddle Points," *Psychological Reports*, 7 (1960), 479–485.

Sakaguchi, M., "Reports on Experimental Games," *Stat. Applied Research JUSE, 7* (1960), 156–165.

Siegel, S., and Goldstein, D. A., "Decision-Making Behavior in a Two-Choice Uncertain Outcome Situation," *Journal of Experimental Psychology, 57,* (1959), 37–42.

Other Matrix Games

Alexander, C. N., Jr., and Weil, H. G., "Players, Persons and Purposes: Situational Meaning and the Prisoner's Dilemma Game," *Sociometry, 32* (1969), 121–144.

Aranoff, D., and Tedeschi, J. T., "Original Stakes and Behavior in the Prisoner's Dilemma Game," *Psychonomic Science, 12* (1968), 79–80.

Arnstein, F., and Feigenbaum, K. D., "Relationship of Three Motives to Choice in the Prisoner's Dilemma," *Psychological Reports, 20* (1967), 751–755.

Birmingham, R. L., "The Prisoner's Dilemma and Mutual Trust: Comment," *Ethics, 79* (1969), 156–158.

Bixenstine, V. E., Chambers, N. and Wilson, K. V., "Asymmetry in Payoff in a Non-Zero-Sum Game," *Journal of Conflict Resolution, 8,* 2, (1964), 151–159.

Bixenstine, V. E., and Gaebelein, J. W., "Strategies of 'Real' Opponents in Eliciting Cooperative Choice in a Prisoner's Dilemma Game," *Journal of Conflict Resolution, 15,* (1971), 157–166.

Bixenstine, V. E. and O'Reilly, E. F., Jr., "Money versus Electric Shock as Payoff in a Prisoner's Dilemma Game," *Psychological Record, 16,* (1966), 251–264.

Bixenstine, V. E., Potash, H. M., and Wilson, K. V., "Effects of Level of Cooperative Choice by the Other Player on Choices in a Prisoner's Dilemma Game: Part I," *Journal of Abnormal and Social Psychology, 66,* (1966), 308–313.; Part II, *JASP, 67,* (1963), 139–147.

Cederblom, D. and Diers, C. J., "Effects of Race and Strategy in the Prisoner's Dilemma," *Journal of Social Psychology, 81,* (1970), 275–276.

Crumbaugh, C. M., and Evans, G. W., "Presentation Format, Other-Person Strategies, and Cooperative Behavior in the Prisoner's Dilemma," *Psychological Reports, 20,* (1967), 895–902.

Dolbear, F. T., Jr., Lave, L. B., Bowman, G., Lieberman, A., Prescott, E., Rueter, F. and Sherman, R., "Collusion in the PD: Number of Strategies," *Journal of Conflict Resolution, 13,* (1969), 252–261.

Evans, G. W., and Crumbaugh, C. M., "Effects of Prisoner's Dilemma Format on Cooperative Behavior," *Journal of Personality and Social Psychology, 3,* (1966), 486–488.

Gahagan, J. P., and Tedeschi, J. T., "Strategy and Credibility of Promises in the Prisoner's Dilemma Game," *Journal of Conflict Resolution, 12,* (1968) 224–234.

Gallo, P. S., Jr., Funk, S. G., and Levine, J. R., "Reward Size, Method of Presentation, and Number of Alternatives in Prisoner's Dilemma Game," *Journal of Personality and Social Psychology, 13,* (1969), 239–244.

Gallo, P. S., Jr., and Winchell, J.D., "Matrix Indices, Large Rewards, and Cooperative Behavior in a Prisoner's Dilemma Game," *Journal of Social Psychology*, *81*, (1970), 235–242.

Geiwitz, J. P., "The Effects of Threats on Prisoner's Dilemma," *Behavioral Science*, *12*, (1967), 232–233.

Gumpert, P., Deutsch, M., and Epstein, Y., "Effects of Incentive Magnitude on Cooperation in the Prisoner's Dilemma Game," *Journal of Personality and Social Psychology*, *11*, (1969), 66–69.

Guyer, M. J., "Response-Dependent Parameter Changes in the Prisoner's Dilemma Game," *Behavioral Science*, *13*, (1968), 205–219.

Harford, T., and Solomon, L., " 'Reformed Sinner' and 'Lapsed Saint' Strategies in the Prisoner's Dilemma Game," *Journal of Conflict Resolution*, *11* (1967), 104–109.

Harris, R. J., "Note on 'Optimal Policies for the Prisoner's Dilemma' ", *Psychological Review*, *76*, (1969), 363–375.

Jones, B., Steele, M., Gahagan, J., and Tedeschi, J., "Matrix Values and Cooperative Behavior in the Prisoner's Dilemma Game," *Journal of Personality and Social Psychology*, *8* (1968), 148–153.

Kanouse, D. E., and Wiest, W. M., "Some Factors Affecting Choice in the Prisoner's Dilemma," *Journal of Conflict Resolution*, *11* (1967), 206–213.

Kelley, H. H., and Stahelski, A. M., "The Inference of Intentions from Moves in the Prisoner's Dilemma Game," *Journal of Experimental Social Psychology*, *6*, (1970), 401–419.

Kershenbaum, B. R., and Komorita, S. S., "Temptation to Defect in the Prisoner's Dilemma Game," *Journal of Personality and Social Psychology*, *16*, (1970), 110–113.

Knox, R. E., and Douglas, R. L., "Trivial Incentives, Marginal Comprehensive, and Dubious Generalizations from Prisoner's Dilemma Studies," *Journal of Personality and Social Psychology*, *20*, (1971), 160–165.

Komorita, S. S., "Cooperative Choice in a Prisoner's Dilemma Game," *Journal of Personality and Social Psychology*, *2* (1965), 741–745.

Lave, L. B., "Factors Affecting Cooperation in the Prisoner's Dilemma," *Behavioral Science*, *10* (1965), 26–38.

Lindskold, S., Gahagan, J., and Tedeschi, J. T., "The Ethical Shift in the Prisoner's Dilemma Game," *Psychonomic Science*, *15* (1969), 303–304.

Marwell, G., Ratcliff, K., and Schmitt, D. R., "Minimizing Differences in a Maximizing Difference Game," *Journal of Personality and Social Psychology*, *12* (1969), 158–183.

Morehous, L. G., "One-Play, Two-Play, Five-Play, Ten-Play Runs of Prisoner's Dilemma," *Journal of Conflict Resolution*, *10* (1966), 354–362.

Morehous, L. G., "Two Motivations for Defection in Prisoner's Dilemma Games," *General Systems*, *11* (1966), 225–228.

Noland, S. J. and Cation, D. W., "Cooperative Behavior among High School Students on the Prisoner's Dilemma Game, "*Psychological Reports*, *24* (1969), 711–718.

Oskamp, S., and Perlman, D., "Factors Affecting Cooperation in a Prisoner's

Dilemma Game," *Journal of Conflict Resolution, 9,* (1965), 359–374.

Plon, M., "Conformité Verbale et Comportementale dans le Cadre d'un 'Dilemme des Prisonniers,' " *Psychologie Française, 12,* (1967), 305–316.

Radlow, R., "An Experimental Study of 'Cooperation' in the Prisoner's Dilemma Game," *Journal of Conflict Resolution, 9,* (1965), 221–227.

Rapoport, Amnon, "Optimal Policies for the Prisoner's Dilemma," *Psychological Dilemma, 74,* (1967), 136–148.

Rapoport, A., "A Note on the 'Index of Cooperation' for Prisoner's Dilemma," *Journal of Conflict Resolution, 11,* (1967), 101–103.

Rapoport, A., and Chammah, A. M., "Sex Differences in Factors Contributing to the Level of Cooperation in the Prisoner's Dilemma Game," *Journal of Personality and Social Psychology, 2* (1965), 831–838.

Rapoport, A., and Chammah, A. M., "The Game of Chicken," *The American Behavioral Scientist, 10* November 1966, 10–27.

Rapoport, A., and Guyer, M., "A Taxonomy of 2 X 2 Games," *General Systems,* 11 (1966), 203–214.

Rapoport, A., and Orwant, C., "Experimental Games: A Review," *Behavioral Science,* 7 (1962), 1–37.

Roberts, F. J., "Conjoint Marital Therapy and the Prisoner's Dilemma," *British Journal of Medical Psychology 44* 1971, 67–74.

Scodel, A., and Minas, J. S., "The Behavior of Prisoners in a Prisoner's Dilemma Game," *Journal of Psychology 50* (1960), 133–138.

Sherman, R., "Individual Attitude toward Risk and Choice between Prisoner's Dilemma Game," *Journal of Psychology, 66* (1967), 291–298.

Shubik, M., "Some Experimental Non-Zero-Sum Games with Lack of Information about the Rules," *Management Science 8,* 2 (1962), 215–234.

Shubik, M., "Games Theory, Behavior, and the Paradox of the Prisoner's Dilemma: Three Solutions," *Journal of Conflict Resolution, 14* (1970), 181–193.

Shubik, M., and Stern, D., "Some Experimental Nonconstant-Sum Games Revisited," Parts I, II and III, New Haven: Yale University; Cowles Foundation for Research in Economics, CFDP No. 236 (1967); CFDP No. 240 (1967); CFDP No. 247 (1969).

Shubik, M., and Wolf, G., "Solution Concepts and Psychological Motivation in Prisoner's Dilemma Games," *Decision Sciences, 5,* 2 (1974), 153–163.

Shubik, M., Wolf, G., and Poon, B., "Perception of Payoff Structure of Opponent's Behavior in Related Matrix Games," *Journal of Conflict Resolution, 18,* 4 (December 1974), 646–656.

Swensson, R. G., "Cooperation in the Prisoner's Dilemma Game: I. The Effects of Asymmetric Payoff Information and Explicit Communication," *Behavioral Science, 12* (1967), 314–322.

Tedeschi, J. T., Aranoff, D., and Gahagan, J. P., "Discrimination of Outcomes in a Prisoner's Dilemma Game," *Psychonomic Science, 11* (1968), 301–302.

Terhune, K. W., "Motives, Situation, and Interpersonal Conflict Within Prisoner's Dilemma," *Journal of Personality and Social Psychology, 8* (1968), 1–24.

Wilson, W., "Reciprocation and Other Techniques for Inducing Cooperation in the Prisoner's Dilemma Game," *Journal of Conflict Resolution*, 15, (1971), 167–195.

Games with Other Formats

Deutsch, M., and Krauss, R. M., "Studies of Interpersonal Bargaining," *Journal of Conflict Resolution*, 6, (1962), 52–76.

Schelling, T. C., "Experimental Games and Bargaining Theory," *World Politics*, 14 (1961), 47–68.

Shubik, M., "A Note on a Simulated Stock Market," *Decision Sciences*, 1 (1970), 129–141.

Shure, G. H. and Meeker, R. J., "Bargaining Processes in Experimental Territorial Conflict Situations," *Peace Research Society (International) Papers*, *XI* (1969), 109–122.

Stone, J. J., "An Experiment in Bargaining Games," *Econometrica* **26** (1958), 286–296.

Three-Person and Larger Experimental Games

Bond, J. R., and Vinacke, W. E., "Coalitions in Mixed-Sex Triads," *Sociometry*, *24* (1961), 61–75.

Caplow, T., "A Theory of Coalitions in the Triad," *American Sociological Review*, *21* (1956), 489–493.

Chertkoff, J. M., "Sociopsychological Theories and Research on Coalition Formation," *in* S. Groennings, E. W. Kelly, and M. Lieserson (Eds.), *The Study of Coalition Behavior*, New York: Holt, Rinehart and Winston, 1970, 297–322.

Gamson, W. A., "An Experimental Test of a Theory of Coalition Formation," *American Sociological Review*, *26* (1961), 565–573.

Horowitz, A. D., "The Competitive Bargaining Set for Cooperative N-Person Games," *Journal of Mathematical Psychology*, *10*, (1973), 265–289.

Horowitz, A. D., and Rapoport, Amnon, "Test of the Kernel and Two Bargaining Set Models in Four- and Five-Person Games," In Anatol Rapoport (Ed.) *Game Theory as a Theory of Conflict Resolution*, Dordrecht, Holland: Reidel, 1974, 161–192.

Kalisch, G. K., Milnor, J. W., Nash, J. F. and Nering, E. D., "Some Experimental N-Person Games," In R. M. Thrall, R. H. Coombs, and R. L. Davis (Eds), *Decision Processes*, New York: Wiley, 1954 301–327.

Kelley, H., and Arrowood, A. J., "Coalitions in the Triad: Critique and Experiment," *Sociometry*, *23* (1960), 231–244.

Laing, J. D., and Morrison, R. J., "Coalitions and Payoffs in Three-Person Sequential Games: Initial Tests of Two Formal Models," *Journal of Mathematical Sociology*, *3* (1973), 3–26.

Laing, J. D., and Morrison, R. J., "Sequential Games of Status," *Behavioral Science*, *19* (1974), 177–196.

Lieberman, B., "I-Trust, A Notion of Trust in Three Person Games and International Affairs," *Journal of Conflict Resolution*, *8* (1964) 271–280.

Lieberman, B., "Experimental Studies of Conflict in Some Two and Three Person Games," *in* F. Massarik and P. Ratoosh (Eds.), *Mathematical Explorations in Behavioral Science*, Homewood, Ill.: R. D. Irwin, 1965, 121–139.

Lieberman, B., "Not an Artifact," *Journal of Conflict Resolution 15*, 1 (1971) 113–120.

Maschler, M., "Playing an N-Person Game: An Experiment," *in* Princeton University; *Recent Advances in Game Theory* (Princeton University Conference 73), 1965), 49–56.

Overstreet, R. E., "Social Exchange in a Three-Person Game," *Journal of Conflict Resolution, 17*, (1972), 109–123.

Philips, J. L., and Nitz, L., "Social Contacts in a Three-Person Political Convention Situation," *Journal of Conflict Resolution, 12*, 2 (1968), 206–214.

Riker, W. H., and Niemi, R. G., "Anonymity and Rationality in the Essential Three-Person Game," *Human Relations, 17*, (1964), 131–141.

Selten, R., "Equal Share Analysis of Characteristic Function Experiments," *in* H. Sauermann (Ed.), *Beitrage Zur Experimentellen Wirtschaftsforschung*, Tubingen: Mohr, 1972, 130–165.

Shubik, M., "Games of Status," *Behavioral Science, 16*, 2 (1971), 117–129.

Shubik, M., "A Three-Person Cooperative Game: Some Experiments with Opinions and Value Systems," Melbourne, Australia: University of Melbourne (July 1973). (Unpublished paper, reported in M. Shubik, *Games for Society, Business and War*, Chapter 11).

Stryker, S., and Psathas, G., "Research on Coalitions in the Triad: Findings, Problems and Strategy," *Sociometry, 23*, (1960), 217–230.

Vinacke, W. E., "The Effect of Cumulative Score on Coalition Formation in Triads with Various Patterns of Internal Power," *American Psychologist, 14* (1959), 381.

Vinacke, W. E., and Arkoff, A., "Experimental Study of Coalitions in the Triad," *American Sociological Review, 22* (1957), 406–415.

Vinacke, W. E., Crowell, D. C., Dien, D., and Young, V., "The Effect of Information about Strategy on a Three-Person Game," *Behavioral Science, 11* (1966), 180–189.

Willis, R. H., "Coalitions in the Tetrad," *Sociometry, 25* (1962), 358–376.

Willis, R. H., and Long, N. J., "An Experimental Simulation of an International Truel," *Behavioral Science, 12*, 1 (1967) 24–32.

Bibliography on Experimental Economics

The attention of the reader is called to the October 1969 issue of the *Review of Economic Studies*, which has a symposium on experimental economics containing six articles.

Alpert, B., "Non-Businessmen as Surrogates for Businessmen in Behavioral Experiments," *The Journal of Business, 40* (1967), 203–207.

Becker, O., and Selten, R., "Experiences with the Management Game SINTO—Market," in H. Sauermann (Ed.), *Contributions to Experimental Economics*, Vol. 2, Tubingen: Mohr, 1970, 136–170.

Carlson, J. A., "The Stability of an Experimental Market with a Supply-

Response Lag," *Southern Economic Journal*, *33*, 3 (January 1967), 305–321.

Castro, B., and Weingarten, K., "Toward Experimental Economics," *Journal of Political Economy*, *78* (May/June 1970), 598–607.

Chamberlin, E. H., "An Experimental Imperfect Market," *Journal of Political Economy*, *56* (April 1948), 95–108.

Conlisk, J., and Watts, H., "A Model for Optimizing Experimental Designs for Estimating Response Surface," Talk presented at American Statistical Association Meeting, Iowa City: University of Iowa, April 1969.

Cummings, L. L., and Harnett, D. L., "Bargaining Behavior in a Symmetric Bargaining Triad," *The Review of Economic Studies*, *36*, 108 (October 1969), 485–501.

Dolbear, F. T., "Individual Choice under Uncertainty: An Experimental Study," *Yale Economic Essays*, 3, (1963), 419–470.

Dolbear, F. T., Jr., Lave, L. A., Bowman, G. Lieberman, A., Prescott, E., Rueter, F., and Sherman, R., "Collusion in Oligopoly: An Experiment on the Effect of Numbers and Information." *Quarterly Journal of Economics*, *82* (February 1968), 240–259.

Dufty, N. F., An Experimental Approach to the Theory of the Firm," *The Economic Record*, *37*, 80 (December 1961), 503–508.

Fouraker, L. E., Shubik, M., and Siegel, S., "Oligopoly Bargaining: The Quantity Adjuster Models," University Park, Pa., Pennsylvania State University (1961) RB #20. Partially reported in Fouraker, L. E., and Siegel, S. *Bargaining Behavior*, New York: McGraw-Hill, 1953.

Frahm, D., and Schrader, L. F., "An Experimental Comparison of Pricing in Two Auction Systems," *American Journal of Agricultural Economics*, *52*, 4 (November 1970), 528–534.

Friedman, J. W., "Individual Behavior in Oligopolistic Markets: An Experimental Study," *Yale Economic Essays*, 3, (1963), 359–417.

Friedman, J. W., "An Experimental Study of Cooperative Duopoly," *Econometrica*, *35* (October 1967), 379–397.

Friedman, J. W., "On Experimental Research in Oligopoly," *The Review of Economic Studies*, *36* (1969), 399–415.

Friedman, J. W., "Equal Profits as a Fair Division," *in* H. Sauermann (Eds.), *Contributions to Experimental Economics*, Vol. 2, Tubingen: Mohr, 1970. 19–32.

Gordon, M. J., Paradis, G. E., and Rorke, C. H., "Experimental Evidence on Alternative Portfolio Decision Rules," *The American Economic Review*, *62*, 1 (1972), 107–118.

Harnett, D. L., and Hamner, W. C., "The Value of Information in Bargaining," *Western Economic Journal*, *11*, 1 (1973), 81–88.

Harnett, D. L. Hughes, D. G., and Cummings, L. L., "Bilateral Monopolistic Bargaining Through an Intermediary," *The Journal of Business*, *41*, (1968) 251–259.

Hoggatt, A. C., "An Experimental Business Game," *Behavioral Science*, 4 (July 1959), 192–203.

Hoggatt, A. C., "Measuring Behavior in Quantity Variation Duopoly Games," *Behavioral Science*, *12*, (1967), 109–121.

Hoggatt, A. C., "Response of Paid Student Subjects to Differential Behaviour of Robots in Bifurcated Duopoly Games," *Review of Economic Studies*, 36 (October 1969), 417–432.

Hoggatt, A. C., "On Economic Experiments in Economics," *In* H. Sauermann (Ed.), *Contributions to Experimental Economics*, Vol. 3, Tubingen: Mohr, 1972, 6–27.

Johnson, H. L., and Cohen, A. M., "Experiments in Behavioral Economics," *Behavioral Science, 12* (1967), 353–372.

Selten, R., "Ein Marktexperiment," *in* H. Sauermann (Ed.), *Contributions to Experimental Economics*, Vol. 2, Tubingen: Mohr, 1970, 33–98.

Shubik, M., "A Note on a Simulated Stock Market," *Decision Sciences, 1* (1970) 129–141.

Shubik, M., "A Trading Model to Avoid Tatonnement Metaphysics," New Haven: Yale University; Cowles Foundation for Research in Economics, CFDP 368 (February 1974).

Shubik, M., and Riese, M., "An Experiment with Ten Duopoly Games and Beat-the-Average Behavior," *in* H. Sauermann (Ed.), *Contributions to Experimental Economics*, Vol. 3, Tubingen: Mohr, 1972, 656–689.

Shubik, M., G. Wolf, and Eisenberg, H., "Some Experiences with an Experimental Oligopoly Business Game," *General Systems, 13*, (1972), 61–75.

Shubik, M., Wolf, G., and Lockhart, S., "An Artificial Player for a Business Market Game," *Simulation and Games 2*, 1 (1971), 27–43.

Simon, J. L., Puig, C. M. and Aschoff, J., "A Duopoly Simulation and Richer Theory: An End to Cournot," *The Review of Economic Studies*, XL, 3 (1973), 353–366.

Smith, V. L., "Experimental Studies of Competitive Market Behavior," *Journal of Political Economy*, 70 (April 1962), 111–137.

Smith, V. L., "Experimental Auction Markets and the Walrasian Hypothesis," *Journal of Political Economy*, 73, (August 1965), 387–393.

Smith, V. L., "Experimental Studies of Discrimination versus Competition in Sealed-Bid Auction Markets," *Journal of Business, 40*, (January 1967), 56–84.

Starbuck, W. H. and Bass, F. M., "An Experimental Study of Risk-Taking and the Value of Information in a New Product Context," *The Journal of Business, 40* (1967), 155–165.

Stech, F. J., and McClintock, C. G., "Cooperative and Competitive Price Bidding in a Duopoly Game," *Behavioral Science, 16* (1971), 545–557.

Stern, D., "Some Notes on Oligopoly Theory and Experiments," *in* M. Shubik (Ed.) *Essays in Mathematical Economics in Honor of Oskar Morgenstern*, Princeton: Princeton University Press, 1967, 255–281.

COMMENTS

Fouraker and Siegel, *Bargaining Behavior*: This work done partially with Harnett and Shubik extended the results on bilateral monopoly and also began the investigation of price variation and quantity variation oligopoly. Specifically, the models considered were duopoly and triopoly under varying conditions of information. In this book the unequal strength models associated with Bowley's price-leadership model of bilateral monopoly were tested. Eight experiments were run controlling for three conditions: information, the form of bidding (single transaction or repeated transactions), and the position of the equal split outcome. Five experiments were run, and it was shown that under single-bid conditions the Bowley point provides a good prediction. However, this solution is weakened as a predictor when repeated bids are allowed. The last chapter, on bilateral monopoly, deals with sources of variability in bargaining. In the experiments on quantity-variation oligopoly done with Shubik, it was felt that three types of players should be distinguished: (1) the simple maximizer, who was concerned with his own results; (2) the rivalist; and (3) the cooperator. It was expected that differential behavior would be found as information conditions were varied. The case corresponding the most closely to the theory of Cournot would be one giving incomplete information, where the individuals know only their own profits and have no information concerning the profits of their competitors. Under complete information, each will be informed about the profits of the others. There are several variants concerning information when there are three, rather than two competitors. These variants were not discussed here or experimented with. There were 130 students participating in the quantity variation duopoly experiments and 66 in the triopoly; 135 students participated in the price variation experiments. Under incomplete information, there were strong results in favor of the Cournot solution for quantity duopoly. The Cournot model was also confirmed for triopolists under incomplete information; as was predicted, the duopoly conditions give rise to more dispersion than does the triopoly. The results on price duopoly and triopoly also confirm the nonequilibrium hypothesis. As was to be expected from theory, the price variation model is essentially more "cutthroat" than the quantity variation model, and this showed up clearly in the triopoly.

Rapoport and Chammah, *Prisoner's Dilemma*; This is a valuable book, which reports on a series of different experiments on the Prisoner's Dilemma game and simultaneously considers several different theoretical approaches to explain observed behavior. It demonstrated the importance of being willing to run many variants of a simple experiment many times. Of particular interest were the clear results of the authors that the size of the numbers in the payoff matrices clearly made a difference to the way in which the players played. In particular, even if the players did not know each other and did not play together, there is evidence that interpersonal comparisons were relevant. Two assumptions frequently made by economists—that individuals behave as if only preference orderings were relevant and as if interpersonal comparisons were not relevant—were not borne out by these experiments.

Rapoport, Guyer, and Gordon, *The 2 X 2 Game*: Rapoport and his colleagues have followed up his original work by an encyclopedic book on experiments on the 2 X 2 matrix game. In this book they not only report on a mountain of work done by themselves, but they also attempt to cover all of the available work of others. In a previous work, Rapoport, and Guyer constructed a taxonomy of 2 X 2 matrix games in which they were able to show that if one considered only the original property of the payoff and if one excluded ties, there are 78 strategically different matrices. In this book experimental results are reported on experiments with all of these matrices.

Results are reported on play in both one-shot matrix games and play in sequential games. There exists a considerable body of literature on the sociopsychological properties of play in matrix games. Chapter 19 deals with "who plays how." The book is 25 chapters in length and is split into four parts, with the first part devoted to explaining the various taxonomies of 2 X 2 games, the game-theoretic approaches, the behavioral theory approaches to play, and the methods and variables employed in these studies. Part II considers the stability of outcome, the class of no-conflict games, the problems involved in competition and cooperation, and, in terms of the experimental game, the roles of preemption and procrastination. Part II covers the effect of payoffs in single- or one-shot play, the effects of payoffs in interated games, the role of the strategy of the other player, conditions on information and communication, and sociopsychological factors. In Part IV the maxmin solution to a zero-sum game is tested, and there is a chapter on the testing of Nash's solution to a cooperative game. There is also a large, annotated bibliography. This is an extremely worthwhile book. Possibly one of the few major criticisms which can be made about it concerns its discussion and treatment of the game-theoretic basis for the work. In fact, there is no decent dynamic theory of games, the phenonema being studied here are, for the most part dynamic. Theory of any type concerning behavior in dynamic, competitive, or cooperative environments is, to say the least, rudimentary. The authors of this book simultaneously appear to give the current developments in game theory too little and too much credit as an aid in interpretation of experimental results. It is my opinion that a little more discussion on the strength and limitations of experimenting with the 2 X 2 matrix game should have been given. In particular, the lack of degrees in freedom in the 2 X 2 game may suppress many important sociopsychological phenomena. In actual life, individuals often complicate situations in order to simplify their solutions. In particular, escrow arrangements, the invention of a whole host of bargaining checks in complex negotiations, and the whole problem of purposely making situations vague and sufficiently complicated so that the degree of conflict of interest is not clearcut are important parts of human conflict resolution. Obviously, one should do simple experiments first. The comments above are not meant to denegrate the work. However, a reader could be helped by comments from the authors concerning how natural or unnatural a social situation is represented by a game. Do the authors believe that generalizations can be drawn with safety, or do they believe that the results show behavior caused by the artifact of the simplicity of the situation?

Sauermann (Ed.), *Contributions to Experimental Economics*: These three volumes provide a handy but somewhat uneven compendium of work in experimental economics and experimental game theory in both Europe and the United States. The articles range from excellent to poor. One might argue that special collections of this variety are not necessary, because the papers could have been published in the economics journals. Unfortunately, in this somewhat imperfect world, articles that cannot be neatly classified as clearly belonging to the central part of a discipline can easily be rejected by the established journals, even if the individual article is of considerable merit. When fashions change, so do the criteria for the acceptability of articles in different professional journals. For those interested in experimental economics and experimental game theory, these volumes are of use.

Schelling, *The Strategy of Conflict*; Shubik, *Game Theory and Related Approaches to Social Behavior; Games for Society, Business and War*: These three books are not primarily concerned with experimental gaming, but are relevant to it. *The Strategy of Conflict* provides many suggestive examples for the experimenter. Although no formal experiments are reported on by Schelling, it is clear that many of his conjectures and assertions can be investigated by experimental means. *Game Theory and Related Approaches to Social Behavior* contains a section surveying the literature on

156

experimental gaming. Chapter 11 of *Games for Society, Business and War* contains surveys of both two-person game experiments and experiments involving three or more players.

Siegel and Fouraker, *Bargaining and Group Decision Making*: This is an excellent brief classic describing some beautiful, simple experiments in economics and psychology. The book consists of five chapters with the first devoted to theoretical formulation, the second describing procedures of the experiment, the third and fourth describing the actual experiments, and the fifth providing an overview and integration. The experiments are on bilateral monopoly. In the first chapter the theoretical work of Bowley, Zeuthen, Cournot, Nash, and several others are discussed. The importance of the role of differential amounts of information is noted, and it is observed that few of the announced theories have taken this into account. The subjects were 116 students at Pennsylvania State University. The procedures under which they were run are described in Chapter 2. The first hypothesis tested for is that the contracts arrived at in bilateral monopoly bargaining would tend to be Pareto optimal. Twelve bargaining pairs were used in experiments to validate this. The results were in good accordance with theory. Further experiments were then made under conditions of incomplete-complete information and incomplete-incomplete information, i.e., when one or both of the bargainers did not have information concerning the other's costs and profits. The hypothesis was that deviation from the Pareto optimum would be least under the conditions of complete-incomplete information and greatest under incomplete-incomplete information. The data tends to confirm the hypothesis. Chapter 4 is devoted to a study of differential payoffs. A hypothesis tested is that the more information available to bargainers, the smaller will be the difference in payoff to each member of any team. The data strongly confirms the hypothesis. Hypotheses suggested by both Fellner and Schelling were examined. A discussion of the role of aspiration levels of the individuals is also given. This book was the recipient of the Monograph Prize of the American Academy of Arts and Sciences for 1959 in the field of the social sciences, and the reason for this award is clear if one reads the book.

Tedeschi, Schlenker, and Bonoma, *Conflict, Power and Games*: In contrast with the other books reviewed here, this is written by social psychologists and psychologists. It contains a very different view of gaming than do the other books and covers a range of sociopsychological literature that is scarcely referred to in the other books. The first four chapters deal with social interaction and experimental games; the sociopsychological components for a theory of conflict; a discussion of compliance to social influence; and a discussion of the exercise of power. They can be of use to the gamer whose background is primarily economic, operations research, or military gaming. The price paid for strength in one area is weakness elsewhere. The game-theoretic coverage is weak. This shows up especially in Chapters 5 and 6, which deal respectively with two-person bargaining behavior in complex games and coalition behavior in n-person groups. The first of these chapters leaves much to be desired, although the literature coverage is both good and a valuable source for the characterization of the differences in the formal structure of the various gaming experiments. It is important to be extremely careful in specifying formal and informal communications structure when games are played more than once or when verbal interchange is permitted. This is not adequately handled here. Even so, this is an informative and useful chapter. Chapter 5, however, on coalition behavior is inadequate. The authors are apparently unaware of a considerable amount of literature on coalitional games with three or more players (see the bibliography in this chapter). They also appear to be unaware of several alternative game-theory cooperative solutions, such as the bargaining set, the nucleolus, the competitive equilibrium point, and stable sets. There is also the possibility that in three-person coalitional games the players may convert the game that is offered in point form into a game of social status.

In spite of the gaps in literature, the coverage of the social-psychological experiments is thorough.

The book ends with a chapter on games as research tools and another chapter on generalizations and applications of the theory of conflict. It is a useful contribution.

Bixenstine, Chambers, and Wilson, "Asymmetry in Payoff in a Non-Zero Sum Game": Many of the experiments with 2 X 2 matrices have been performed with symmetric matrices. In this experiment a highly nonsymmetric matrix was used as is shown below:

	1	2
1	3, 2	2, 2
2	10, 1	1, 1

It may be observed that the payoff to the second player is independent of his choice. However, he has considerable power in determining the payoff to the first player.

The experimental subjects were 32 female and 32 male students, for the most part, from a general psychology course. An ethicality measure was administered to all subjects before they were paired. The prediction that individuals would be more cooperative in the column role than in the row role (Player 1) was not borne out. However, male subjects appeared to be significantly more cooperative than women subjects. The games were played for a length of 80 trials, after which each subject was then required to play in the other position; i.e., the row player (Player 1) became the column player (Player 2)) and then played for another 80 trials. This is a quite complicated matrix to play sequentially, and the outcomes clearly depend not only upon the individuals' innate level of cooperation but also upon their ability to signal to each other. Before one can go very far in interpreting the experimental results in a situation such as this, it is necessary to have at least some conjecture and analysis of communication from the viewpoint of both learning and teaching. For example, a not unreasonable strategy for Column would be to alternate between 1 and 2 with complete regularity, thus signalling to Row that he should alternate between 2 and 1. In this manner Row obtains 10 on the odd periods against 1 for Column, but Column obtains 2 on the even periods, giving Row 2 on those periods.

Chamberlin, "An Experimental Imperfect Market": This is the first experimental market game reported on in the literature. The subjects were students in a class on economic theory. They participated as buyers and sellers in a market. Approximately half were buyers and the other half sellers. Cards were passed out with a B or an S—for buyer or seller. Next to the letter was a number, for example, "B-36" indicates that the buyer would be willing to pay only as high as 36 in purchasing the one unit of commodity he wishes to purchase. Similarly, "S-20" indicates that the seller wishes to sell his one unit at a price no lower than 20. All buyers and sellers were in the market for only one unit. They were permitted to wander around seeking to conclude bargains with each other, but hiding the information they had on their cards. After a contract was made, they surrendered the tickets at a desk.

In the 46 experiments the actual volume of sales was higher than the competitive equilibrium amount 42 times and the same 4 times. The average price was higher than the equilibrium price 7 times and lower 39 times.

This work, with a two-sided market, contrasts with the later developments in the study of oligopolistic markets but is closely linked with the subsequent work of Vernon Smith and Shubik.

Chertkoff, "Social Psychological Theories and Research on Coalition Formation": This is a good review article of socio-psychological literature on three person coalitions. It discusses and covers the conjectures of Caplow and the work by Gamson and others on how coalitions may form in the triad. A more detailed review of this article together with the game theory literature on three-person, experimental, cooperative games is given in reasonable detail in Shubik, *Games for Society, Business and War*, Chapter 11. For those interested in experimentation with the triad, this is an extremely useful article to provide an overview.

Friedman, "An Experimental Study of Cooperative Duopoly": In this experiment, a set of complete information duopoly games were run in which the subjects were permitted to communicate by means of a written message before each decision. The experiment consisted of three replications each using 6 different subjects for a total of 18 subjects. The information was displayed by means of payoff matrices of size 30 X 30. The payoff functions were cubics, and various parameter settings were used to differentiate the nine treatments. The experiment was distinguished by having some where the payoffs were known to be money and others in which the ratio of conversion from points to money for the competitor was unknown. The players were hired and paid. Three questions were investigated: The first concerned the probability that a pair of subjects would agree on the prices to charge; the second the probability that the decision would be Pareto optimal; and the third, whether or not the Nash solution, the joint maximum, or the equal profits point on the Pareto optimal surface would be the best predictor for the outcome chosen. if it were Pareto optimal.

The results showed that in over three-quarters of all the periods agreements were reached and among the agreements over three-quarters were at Pareto optimal points. Furthermore, the evidence tended to give some support to the Nash point among the outcomes chosen on the Pareto optimal surface.

Hoggatt, "Measuring Behavior in Quantity Variation Duopoly": In this experiment Hoggatt had his subjects play in two games simultaneously. They were informed that they were selling the two completely isolated and independent markets, each with a distinct competitor. The competitors were robot players, each parameterized differently so that one showed a high and the other showed no degree of cooperation. The main hypothesis of Hoggatt was that the higher level of cooperation from a robot would evince a higher level of cooperation from the given player. This was substantiated. In a subsequent experiment, "Behavior of Robots in Bifurcated Duopoly Games," this experiment has been considerably elaborated.

Lieberman, "Human Behavior in a Strictly Determined 3 × 3 Matrix Game": Fifteen pairs of subjects played in the 3 × 3 matrix game shown below for 200 plays each, for a reward of a penny a point. During the last 10 trials around 90 percent of the population selected their optimal strategies. It appears as if boredom and low stakes accounted for the deviation.

	1	2	3 ↓	min
1	15	0	-2	-2
2	0	-15	-1	-15
→ 3	1	2	0	0 max
max	15	2	0 min	

The maxmin strategies suggested from two-person constant sum game theory can be regarded as a normative solution concept. This solution tells the cautious player how much he can absolutely guarantee for himself regardless of the behavior of his opponent. There are two clear levels of difficulties encountered in experiments with constant-sum, two-person games. Some games have saddle points and others call for mixed strategies. It is much harder to make out a case justifying the utilization of a mixed strategy than it is to justify the employment of saddlepoint strategy. This is discussed further in Chapter 21 of the book on the 2 × 2 game by Rapoport et al. The Lieberman experiment was confined to the saddlepoint situation.

Maschler, "Playing an N-Person Game, An Experiment": This article is different from the one by Chertkoff. Here a game theorist describes an experiment with 38 pupils in 123 plays of three-person and four-person games where the author and colleagues have constructed a new solution theory known as the bargaining set in order to explain the behavior observed. The nature of the approach is highly different from that of the social psychologists. The benefits from bringing the two more closely together can be seen from reading this paper together with the paper by Chertkoff. An excellent summary of these and other results is provided by Selten, already referred to in the bibliography "Equal Share Analysis of Characteristic Functions."

Morehaus, "One Play, Two Play, Five Play and Ten Play Runs of a Prisoner's Dilemma": The subjects in this experiment were 200 undergraduate women enrolled in a beginning psychology course. Every pair of subjects played a one trial game, a two-trial game, a five-trial game and a ten-trial game.

This paper is noted as an example of a small matrix-game experiment that can be performed relatively easily and that serves to build up an inventory of hundreds of experiments needed to provide a basis for the substantiation of assertions and conjectures concerning behavior in even extremely simple dynamic games with conflicting interest.

Shubik, "Some Experimental Non Zero Sum Games with Lack of Information about the Rules": Several groups of students played in six 2×2 sequential matrix games where each student was informed only of his own payoff, but not those of his competitor or partner. The students were not aware of the period of play in each game. Scores were cumulative from period to period and game to game. Four different static game-theory solutions were considered and compared with the actual behavior of the students. It was observed that in a game with lack of information and low communication, the noncooperative equilibrium appeared to give the best results, but where the noncooperative equilibrium point had other properties—such as uniqueness, Pareto optimality, or symmetry—then its value as a predictor was considerably enhanced. In subsequent replications of the experiment students were asked to guess the ordering on the payoffs in their competitor's matrix. It is of interest to note that the Prisoner's Dilemma matrix was the hardest for the students to guess.

Shubik, Wolf, and Eisenberg, "Some Experiences with an Experimental Oligopoly Business Game": This is a report of several experiments run over many years with a relatively complex business game. The players were required to make several decisions. These included price, production and advertising. Hence, the decision structure of the game was relatively complicated. In contrast with most of the other experimental work that has been noted in the bibliography, the information provided to the players came in the form of more-or-less standard accounting information together with some com-

162

petitive market information. Games were played with the number of competitors being two, three, five, seven, and ten. In a different series there were also some four-person games.

The data generated were examined with respect to the static theories of cooperative monopolistic behavior, the noncooperative equilibrium and the competitive equilibrium. In part the games with many players were played in a "pseudocompetitive mode." In other words, although the individual saw competitive data from one, two, four, or nine other players, the markets were in fact isolated and there was no actual structural interaction. In the pseudomonopoly situation, price rose slowly but did not attain full monopoly level. In the competitive situation, price which was initially selected to be closer to monopolistic than the competitive price declined within 12 periods to the competitive level, although it did not evince any stability. Cyclical behavior was observed and many of the players indulged in cutthroat competition. The variability was greatest for the duopolistic cases.

Considerably more variation was encountered with the advertising variable. In particular, the game was parameterized in such a manner that joint maximum advertising was in fact 0 whereas beat-the-average advertising in all instances was around $5 million. The noncooperative equilibrium in advertising converged from 0 at monopoly to slightly under the beat-the-average value for 10 players. Surprisingly, the experimental data went in the other direction. The variance in advertising was the greatest for duopoly and the 10-person market. However, as the numbers increased, the average level of advertising dropped so that it was the least competitive for 10 players. We suspect that whereas even a naive player has some feeling about the relationship among price, profit, and sales, the understanding of the effect of advertising is far more subtle. Especially in this particular market structure, cooperative behavior calls for less advertising yet at the same time when the number of players is many, a simple hypothesis that "My advertising has little effect on the market, hence I might as well save the money," provides a simple reason to do the right thing.

Shubik, "A Three Person Cooperative Game: Some Experiments with Opinions and a Value System": This experiment, which is reported on more fully in *Games for Society, Business and War*, involved a three-person game presented in coalitional form. Subjects were given the characteristic function of the game as well as a representation of the set of imputations. Each individual was asked separately what he or she thought should be the outcome to a play of the game. Each was also asked to justify in writing his selection. In one run of the game, the subjects were asked to specify both how they thought the game should be played and how the game would be played if they were required to be randomly matched with other respondents present. The respondents were, for the most part, economics faculty members, although in some case they were mathematicians, and in another case undergraduates. Their choices were examined and compared with the predictions of various cooperative game solution theories. These solution theories included the core, the value, the nucleolus, the kernel, the bargaining set, the competitive equilibrium, the point of symmetry, and the stable sets. There was considerable bias toward selecting the point of symmetry. This was backed up with verbal justification of the egalitarian solution. Setting aside the equal-division point, all but one point selected were within the core. Among those points within the core, the solutions selected were a significant number of times near the nucleolus, the value, and the competitive equilibrium.

Smith, "Experimental Study of Competitive Market Behavior": The Smith experiments divided the group into buyers and sellers, each receiving a card containing either the highest price for which he should be willing to buy or the lowest price for which he should be willing to sell a single unit. Each experiment was conducted over a sequence of trading periods, during which buyers and sellers could make bids and offers. During each period the market was regenerated, i.e., new supplies and demands were introduced to produce the same overall market. The experimental subjects were sophomore and junior engineering, economics, and business majors, except in one experiment where a graduate class in economic theory was used. The tentative conclusions from this set of experiments were:

1. Even when numbers are small there are strong tendencies for a supply and demand competitive equilibrium to emerge as long as there is a prohibition of collusion and a dissemination of all bids and offers.
2. Changes in conditions of supply and demand caused changes in the volume of transactions and level of contract price in a manner that corresponds reasonably well with competitive price theory.
3. There is some indication that a prediction of static equilibrium in a competitive market requires knowledge of the

164

shapes of supply and demand schedules as well as the intersections of these schedules.

Perhaps the most important observation from this paper is that the market adjustment mechanism of the so-called Walrasian Hypothesis does not seem to be confirmed. A better hypothesis is that the speed of contract price adjustment is related to the excess of buyer-plus-seller "virtual" rents over the equilibrium buyer-plus-seller rents. Smith investigates this point in a later paper on "Experimental Auction Markets and the Walrasian Hypothesis."

Stone, "An Experiment in Bargaining Games;" In this simple experiment a group of students were presented with 34 diagrams of the variety shown in Figure 9.1. The students did not see their opponent or know who he was. Player 1 was requested to select a horizontal line with the understanding that Player 2 would select a vertical line. If the intersection fell within the area enclosed by the axes and the boundaries of the bargaining area, the players would be paid up according to the coordinates of the intersection of the two lines they had chosen. If the point were to fall outside of the boundary, the players would receive money. Of the 30 diagrams which constituted the experiment (4 others were practice diagrams) there were 13 pairs of diagrams and 2 single diagrams. The property of a pair was that the roles of Player 1 and Player 2 were interchanged. but it would be possible to see how the students' viewpoints changed with his role in the game.

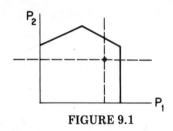

FIGURE 9.1

The experiment can be viewed as an investigation to test both the Zeuthen-Nash theory of bargaining and two-person cooperative games and to test the effectiveness of salient points. Unfortunately, out of 135 players, only 52 provided acceptable data. Even so, neither saliency nor the Zeuthen-Nash theory provided a reasonable explanation of behavior. It would be desirable to see an experiment of this variety performed again with a larger sample and further design.

165

10 Gaming in Political Science, International Relations and for Military Purposes: A Literature Guide

There are many types of gaming and many different purposes served by the gaming noted in this chapter. In particular, political science gaming certainly encompasses more than international relations, although it is in international relations that the activity has been the greatest. For example, there are some games at the level of both playing and formal theory that involve the study of voting behavior and election procedures. The international relations games merge imperceptibly into the military games. They, however, are primarily at the level of strategic gaming. There is a completely different art form in military gaming, essentially concerned with tactical matters and addressed to purely specific military problems without much, if any, emphasis on the political or international-relations consequences.

The military gaming community consists for the most part of "hard gamers" or simulators and model builders, the major exception being those concerned with political military exercises and those who run a certain amount of free-form gaming for other purposes. The interested reader can find adequate descriptions in the books of McHugh, Wilson, and Shubik on the different types of military games.

USAGE

The main users of political and international relations games are the universities at both the undergraduate and graduate level. There is an intermixture between teaching and experimentation. In contrast with the relatively sparse experiments of the game theorist and the social psychologist working with 2 X 2 matrix games or other highly formal structures, the international relations experimenters tend to mix their work heavily with teaching as well as experimentation. For this reason the political science and international relations experimental work has been included in this chapter rather than relegated to Chapter 9.

The political military exercise has been played at the Pentagon, at the Rand Corporation, M.I.T., among members of NATO, and elsewhere. It was possibly considerably more in vogue during the mid-1960s than it is now. Among the more important professionals involved in this type of gaming are, or were, Goldhamer, Bloomfield, Schelling, and Wohlstetter.

Since the creation of the political military exercise and the construction of INS, the major activity in I.R. gaming has been with somewhat free-form models, at least to the point that the programs were not fully computerized and that live players have been used rather than utilizing pure simulations or mathematical models. There has, however, been over the past few years an interest and a tendency to consider strict simulations and mathematical models as tools for the development of theory, data gathering, and empirical methods in political science and international relations. Among those more interested in formal models and simulations are Alker, de Sola Pool, Brunner and Brewer, Deutsch, Kaplan, Snyder, and several others. Their works are not discussed in any detail in the remainder of this chapter. An adequate coverage of this topic would require virtually another whole book.

The users of military gaming form an extremely different community. They include the Pentagon, the war colleges, the military academies, institutions such as Rand, RAC, the Stanford Research Institute, and many large corporations, which at one time or another have contracted for military contracts.

Although popular articles and public information have tended to describe the more newsworthy and science-fiction-like activities of the political military exercises which have the trappings of high-level decision-making, most of the military gaming has been at the tactical level directed to specific problems. The work has primarily employed formal structures, mathematical models, and simulations.

The major uses of military gaming have been for teaching and training, and operational purposes. The operational purposes can be further broken down into technical evaluation, doctrinal evaluation, and force-structure analysis.

Whereas in the study of political and international-relations games, the personnel involved are generally those with a political science or other behavioral science background, in military gaming there is a preponderance of individuals with backgrounds in engineering, mathematics, game theory, and physics.

Not only are there military personnel involved in the serious business of war gaming but there is also a considerable group of amateur war-gamers whose contribution should be no means be ignored.

A BIBLIOGRAPHY ON POLITICAL SCIENCE
AND INTERNATIONAL RELATIONS

Books

Boorman, S. A., *The Protracted Game: A Wei-Ch'i Interpretation of Maoist Revolutionary Strategy*, London: Oxford, 1969.

Guetzkow, H. (Ed.), *Simulation in Social Science: Readings*, Englewood Cliffs, N.J.: Prentice-Hall, 1962.

Guetzkow, H., Alger, C. F., Brody, R., Noel, R. D., and Snyder, R. C., *Simulation in International Relations: Developments for Research and Teaching*, Englewood Cliffs, N.J.: Prentice-Hall, 1963.

Iklé, F. C., *How Nations Negotiate*, New York: Harper & Row, 1964.

Laponce, J. A. and Smoker, P. (Eds.), *Experimentation and Simulation in Political Science*, Toronto: University of Toronto, 1972.

Quade, E. S., and Boucher, W. I., *Systems Analysis and Policy Planning: Applications in Defense*, New York: American Elsevier, 1968.

Rapoport, A., *Fights, Games and Debates*, Ann Arbor, Mich.: University of Michigan Press, 1960.

Raser, J. R., *Simulation and Society: An Exploration of Scientific Gaming*, Boston: Allyn & Bacon, 1969.

Said, A. A. (Ed.), *Theory of International Relations: The Crisis of Relevance*, Englewood Cliffs, N.J.: Prentice-Hall, 1968.

Schelling, T. C., *The Strategy of Conflict*, Cambridge, Mass.: Harvard University Press, 1960.

Shubik, M., *Game Theory and Related Approaches to Social Behavior*, New York: Wiley, 1964.

Shubik, M., *Games for Society, Business and War*, Amsterdam: Elsevier, 1975.

Tanter, R., *Modelling and Managing International Conflict*, Beverly Hills, Calif.: Sage, 1974.

Wilson, A., *The Bomb and the Computer*, New York: Delacorte, 1968, and London: Barrie and Rockliff, 1968.

Articles

Alker, H. R., and Brunner, "Simulating International Conflict: A Comparison of Three Approaches," *ISQ*, *13* (March 1969), 1, 70–110.

Allison, G. T., "Conceptual Models and the Cuban Missile Crisis," *APSR, 63* (September 1969), 689-718.

Banks, M. H., Groom, A. J. R., and Oppenheim, A. M., "Gaming and Simulation in International Relations," *Political Studies, 16* (February 1968), 1-17.

Barringer, R., and Whaley, B., "The M.I.T. Political Military Gaming Experience," *Orbis, 9,* 2 (Summer 1965), 437-458.

Benson, O., "A Simple Diplomatic Game," *in* J. N. Rosenau (Ed.), *International Politics and Foreign Policy,"* New York: Free Press, 1961, 504-511.

Benson, O., "Simulation of International Relations and Diplomacy," *in* H. Borko (Ed.), *Computer Applications in the Behavioral Sciences,* Englewood Cliffs, N.J.: Prentice-Hall, 1962, 574-595.

Bloomfield, L. P., "Political Gaming," *U.S. Naval Institute Proceedings, 86* (1960), 57-64.

Bloomfield, L. P., "Computer and Policy Making: The CASCON Experiment," *Journal of Conflict Resolution, 15* (1971), 33-53.

Bloomfield, L. P., and Padelford, N. J., "Three Experiments in Political Gaming," *APSR, 53* (December 1959), 1105-1115.

Bloomfield, L. P., and Whaley, B., "POLEX: The Political Military Exercise," *The Military Review: Professional Journal of the U.S. Army* (November 1965), 65-71.

Bonham, M. G., "Simulating International Disarmament Negotiations," *Journal of Conflict Resolution, 15,* 3 (September 1971), 299-315.

Boguslaw, R., and Glick, E. B., "A Simulation Vehicle for Studying National Policy Formation in a Less Armed World," *Behavioral Science, 2* (1966), 43-61.

Boulding, K., "Is Peace Researchable?" *Background, 5* (1963), 70-77.

Brody, R. A., "Some Systemic Effects of the Spread of Nuclear Weapons Technology: A Study Through Simulation of a Multi-Nuclear Future," *Journal of Conflict Resolution, 7,* 4 (December 1953), 663-753.

Brody, R. A., "The Uses of Simulation in International Relations Research," *in* J. L. Bernd (Ed.), *Mathematical Applications in Political Science,* Vol. 4, Charlottesville, Va.: University of Virginia Press, 1969, 4-21.

Brown, S., "Escalation in the Game World and the Real World," Santa Monica: The Rand Corporation, D-12567-PR (1967).

Browning, R. P., "Computer Programs as Theories of Political Processes," *Journal of Politics, 24* (August 1962), 562-582.

Bull, H., "International Theory: The Case for the Classical Approach," *World Politics, 18* (1966), 361-377.

Burns, A. L., "International Theory and Historical Explanation," *History and Theory (The Hague), 1,* 1 (December 1960), 55-74.

Guetzkow, H. "Some Correspondences between Simulations and 'Realities in International Relations,' " *In* M. A. Kaplan (Ed.), *New Approaches to International Relations,* New York: St. Martin's Press, 1967, 202-269.

Halperin, M., "Why Bureaucrats Play Games," *Foreign Policy, 2* (Spring 1971), 70-90.

Hermann, C. F., "Validation Problems in Games and Simulations with Special Reference to Models of International Politics," *Behavioral Science, 12* (May 1967). 216-231.

169

Hermann, C. F., and Hermann, M. G., "An Attempt to simulate the outbreak of of World War I," *APSR, 61* (June 1967), 400–416.

Kaysen, C., "The Computer that Printed out W*O*L*F" *Foreign Affairs, 50,* 4 (July 1972), 660–668.

Kegley, C. W., Jr., "Simulation Review: PRINCE," *Simulation and Games, 4,* 1 (March 1973), 110–114.

Levine, R. A., "Crisis Games for Adults," Santa Monica, Calif.: The Rand Corporation 12703-ISA (August 1964).

Licklider, R. E., "Simulation and the Private Nuclear Strategists," *Simulation and Games, 2,* 2 (June 1971), 163–171.

Macoby, M., "Social Psychology of Deterrence," *Bulletin of the Atomic Scientists, 17* (1961), 278–281.

Mathematica Corporation Review of the TEMPER Model. Prepared by M. Balinski, K. Knorr, O. Morgenstern, F. Sand, and M. Shubik. Princeton, N.J.: Mathematica Corp. (1966) mimeo.

McRae, V. V., "Gaming as a Military Research Procedure," *in* I. de Sola Pool (Ed.), *Social Science Research and National Security,* Washington, D.C.: Smithsonian Institution, March 1963, 188–224.

Modelski, G., "Simulation, 'Realities,' and International Relations Theory," *Simulation and Games, 1,* 2 (June 1970), 111–134.

Noël, R. C., "Evolution of the INS," *in* H. Guetzkow et al., *Simulation in International Relations: Developments for Research and Teaching,* Englewood Cliffs, N.J.: Prentice-Hall, 1963, 69–102.

Noël, R. C., "The POLIS Laboratory," *American Behavioral Scientist, 12* (1969) 30–35.

Pilisuk, M., et al., "War Hawks and Peace Doves: Alternative Resolutions of Experimental Conflicts," *Journal of Conflict Resolution, 9* (1968), 491–508.

Pool, I. de Sola, and Kessler, A., "The Kaiser, the Tsar and the Computer," *American Behavioral Scientist, 8,* 1 (1965), 31–38.

Powell, C. A., "Simulation: The Anatomy of a Fad: A Critique and a Suggestion with Respect to Its Use in the Study of International Conflict," *Acta Politica, 4* (1969), 299–330.

Rapoport, A., "Lewis. F. Richardson's Mathematical Theory of War," *Journal of Conflict Resolution, 1* (1957), 249–299.

Raser, J. R., and Crow, W. J., "A Simulation Study of Deterrence Theories," *in* D. Pruitt and R. Snyder (Eds.), *Theory and Research on the Causes of War,* Englewood Cliffs, N.J.: Prentice-Hall, 1969, 136–146.

Robinson, J. A., Hermann, C. F., and Hermann, M. G., "Search under Crisis in Political Gaming and Simulation," *in* D. Pruitt and R. Snyder (Eds.) *Theory and Research on the Causes of War,* Englewood Cliffs, N.J.: Prentice-Hall, 1969, 80–94.

Robinson, T. W., "Game Theory and Politics, Recent Soviet Views," Santa Monica, Calif.: The Rand Corporation, RM-5389-PR (May 1970).

Rosenau, J. T., "Games International Relations Scholars Play," *Journal of International Affairs, 21* (1967), 293–303.

Sachs, S. M., "The Uses and Limits of Simulation Models in Teaching Social Science and History," *The Social Studies, 61,* 4 (April 1970), 163–167.

170

Schelling, T. C., "Experimental Games and Bargaining Theory," *World Politics,* *14,* 1 (1961), 47–68.

Schelling, T. C., "War without Pain, and Other Models," *World Politics, 15* (1963), 465–485.

Schwartz, D. C., "Problems in Political Gaming," *Orbis, 9* (1965), 677–693.

Shubik, M., and Brewer, G. D., "Methodological Advances in Political Gaming: The One-Person, Computer Interactive, Quasi-Rigid Rule Game," *Simulation and Games, 3,* 3 (1972), 329–348.

Singer, J. D., "Data-Making in International Relations," *Behavioral Science, 10* (1965), 68–80.

Singer, J. D., and Hinomoto, H., "Inspecting for Weapons Production: A Modest Computer Simulation," *JPR, 1* (1965), 18–38.

Speier, H., "Political Games and Scenarios," *Public Opinion Quarterly, 25* (1961), 426–427.

Sullivan, D. G., "INS: A Review of its Premises," *in* M. Inbar and C. S. Stoll (eds.), *Simulation and Gaming in Social Science,* New York: Free Press, 1972, 111–124.

Tanter, R., "The Policy Relevance of Models in World Politics," *Journal of Conflict Resolution, 16,* 4 (1972), 555–583.

Verba, S., "Simulation, Reality and Theory in International Relations, *World Politics, 16,* 3 (1964), 490–519.

Waltz, K. N., "Realities, Assumptions and Simulations," *in* W. D. Coplin (Ed.) *Simulation in the Study of Politics,* Chicago: Markham, 1968, 105–102.

Wohlstetter, A., "Sin and Games in America," *in* M. Shubik (Ed.), *Game Theory and Related Approaches to Social Behavior,* New York: Wiley, 1964, 209–225.

Zinnes, D. A., "A Comparison of Hostile State Behavior in Simulated and Historical Data," *World Politics, 18,* 3 (1966), 474–502.

COMMENTARY

Books

Boorman, *The Protracted Game:* This is an interesting and enjoyable small book that makes a direct analogy between the ancient Chinese game of Wei-Ch'i (also known in Japan as the "Game of Go") and the Chinese Communist takeover. The Go Game is played on the map of China. The game of Wei-Chi'i stresses the concepts of encirclement and extremely tenuous control over large areas of territory. In contrast with the game of chess, the situation is simultaneously more fluid and more static and offers a different range of subtleties than chess. If one does not overdo the analogy with the actual political situation, this can be an enjoyable and instructive book.

Guetzkow, et al., *Simulation in International Relations:* This is the classic book in the use of gaming in international relations. Although the work had its roots in the gaming development at Rand, the purposes of the games played there differed considerably from the goals indicated here, which are the improvement of the construction of theory and the teaching of international relations. As the book is a compilation, it suffers somewhat from the usual difficulties of having many authors; it does not quite hold together as a unit. Richard Snyder's essay, "Some Perspectives on the Use of Experimental Techniques in the Study of International Relations" (p. 1–23), presents a survey of trends in simulation and gaming and relevant literature up to 1962. The essay by Guetzkow on "A Use of Simulation in the Study of Internation Relations" gives a history of INS, together with the cases for and against using this type of game. The two essays by Noël, "Internation Simulation Participant's Manual," and "Evolution of the Internation Simulation," give a description of player roles, events, sequencing, and the general mechanics of the game. Both Guetzkow and Noël, who are two of its principal designers, indicate that to a great degree the game stands as a testimony to our lack of scientific knowledge about international relations. As such, they stress that one should have grave reservations about using INS for operational purposes. Alger, "The Use of the Internation Simulation in Undergraduate Teaching," appears to downplay the reservations of his colleagues. Brody, "Varieties of Simulations in International Relations Research," catalogues and discusses several of the more-prominent political diplomatic military games and simulations. A reasonably useful selected bibliography is also supplied.

Iklé, *How Nations Negotiate:* This book is not directly concerned with gaming. It is a careful and perceptive study which outlines many of the variables and indicates the conceptual difficulties in trying to construct models of international negotiations processes. It is of considerable value to a would-be IR gamer to read a book such as this in order to compare his game with the model of the process outlined here. Parenthetically, it must be noted that in virtually all of the work on gaming in IR and on diplomatic and military games, there appears to have been a lack of concern for and a lack of stress on the role of bureaucracy. A general danger that appears to be present in most of the games is that they seem to indicate that once somebody has made up his mind and decided to do something, in fact this gets translated into action. Perhaps one of the better preparatory games for bureaucratic life might be one in which very little happens after the decision, or if it does happen, then it may be extremely distorted and a highly modified version of what the decision-maker had in mind.

Quade and Boucher (Eds.), *Systems Analysis and Policy Planning: Applications in Defense:* This, together with an earlier book edited by Quade on the analysis for military decisions, serves as a highly useful reference for operations research, mathematical, and economic applications to defense planning. The reference to this book is given here as well as in the military classification because it and its companion volume can serve to introduce the political scientist or other behavioral scientist to the problems in model-building and quantitative analysis.

Rapoport, *Fights, Games and Debates:* This well written and instructive book is of use to model-builders of every variety, as it lays stress on the importance of understanding the relationship between a model and the context of the problem being modeled. This is one of several books in which the scope and the difficulties in using game theory is discussed.

Schelling, *The Strategy of Conflict:* This is an important, extremely good, and simultaneously somewhat bad book. It contains an important and imaginative critique of game theory models, yet at the same time provides many instructive and intuitive examples for the application of game theory. Schelling's criticism is well-taken. The mathematical theory of games has not been of great use in studying problems where negotiations and verbal interchanges are of key importance. He identifies the key area of weakness—that is in the modeling and treatment of communication in games played in the extensive form (see Chapter 2 of this book). Although the work is insightful and the criticisms are to a great extent relevant, no alternative body of theory is offered. Schelling's work shows both the value and the limitations of simple 2 × 2 matrix games to provide examples and serve as analogies to strategic problems.

Wilson, *The Bomb and the Computer:* This is a well-written book by a journalist who is Defense and Aviation Correspondent of *The Observer* in Great Britain. He has interviewed many of the major proponents of gaming in most of the active institutions. As a nontechnical introduction to what war gaming is about and to some extent what gaming in international relations is about, this is a useful, nontechnical book.

Articles

Barringer and Whaley, "The M.I.T. Political Military Gaming Experience": This article describes a study made of the participants in a series of political military games conducted at M.I.T.'s Center for International Studies since 1958. The respondents were asked their opinions about the uses of gaming as a technique for teaching and training adjunctive to policy planning and as a research tool. They noted that prior participation in a political-military exercise increased both the quantity and quality of policy alternatives they perceived. This article provides a good appraisal of games and their uses as perceived by the participants.

Levine, "Crisis Games for Adults": This is one of a series of papers written by Levine, Schelling, Jones, and others in an internal debate at Rand about the usefulness of the political-military exercise; the whole series is worth reading. It is a pity that they were not put together and published in a more easily obtainable form, although they can still be obtained from the Rand Corporation.

"Mathematica Corporation Review of the TEMPER Model": The TEMPER model serves as a useful bad example of how *ad hoc* modeling with little empirical background and no theory can chew up hundreds of thousands of dollars, proliferate models, and produce very little. The paper referred to here contains five critical reviews of the model dealing with (1) the actual model in the simulation, its purported purposes and uses; (2) the political inputs in the structure; (3) the economic inputs; (4) the modeling of bargaining, negotiation, and strategic decision-making; and (5) the data base. The reviewers conclude that the model was never viable as an instrument of research. However, it is my belief that it is worth studying and reading the critiques of this model, as it can stand to serve as an extreme example of how to not go about constructing a large-scale simulation.

Wohlstetter, "Sin and Games in America:" A perceptive discussion of the confusion, misunderstanding, and misapplications of game theory to international affairs is presented. This is a highly documented and well written piece and provides a good example of the "two cultures" difficulty encountered when literary and analytical modes are mixed.

A BIBLIOGRAPHY ON MILITARY GAMING

Books

Featherstone, D. F., *Naval War Games*, London: Stanley Paul, 1965; *War Games*, 1965; *Air War Games*, 1966.

Featherstone, D. F., *War Game Campaigns*, London: Stanley Paul, 1970.

Hausrath, A. H., *Venture Simulation in War, Business and Politics*, New York: McGraw-Hill, 1971.

Livermore, W. R., *The American Kriegspiel: A Game for Practicing the Art of War upon a Topographical Map*, Boston: W. B. Clarke (2nd. ed.) 1898.

McHugh, F. J., *Manual of War Games*, Newport, R. I.: U. S. Naval War College, 1960.

McHugh, F. J., *Fundamentals of War Gaming*, Newport, R.I.: U.S. Naval War College (3rd. ed.), 1966.

Morschauser, J., *How to Play War Games in Miniature*, New York: Walker, 1962.

Overholt, J. (Ed.), *First War Gaming Symposium Proceedings*, Washington, D.C.: Washington Operations Research Council, 1961.

Phillips, T. R., Brig. Gen. (Ed.), *Roots of Strategy*, Harrisburg, Pa.: Military Service Publishing Co., March 1955.

Quade, E. S., *Analysis for Military Decisions*, Santa Monica: The Rand Corporation, R-387-PR, 1964.

Quade, E. S., and Boucher, W. I. (Eds.), *Systems Analysis and Policy Planning*, New York: American Elsevier, 1968.

Richardson, L. F., *Arms and Insecurity*, Chicago: Quadrangle Books, 1960.

Richardson, L. F., *Statistics of Deadly Quarrels*, Chicago: Quadrangle Books, 1960.

Sayre, F., *Map Maneuvers*, Fort Leavenworth, Kan.: U.S. Army Staff College Press, 1908.

Shubik, M. S., and Brewer, G. D., *Models, Simulations and Games: A Survey*, Santa Monica, Calif.: The Rand Corporation, R-1060-ARPA/RC, May 1972.

Ulhaner, H. E. (Ed.), *Psychological Research in National Defense Today* (Proceedings: American Psychological Association, Military Psychology Division, 1964), Washington, D.C.: U.S. Army Behavioral Science Research Laboratory, Tech. Rep. S-1, June 1967.

U.S. Army Strategy and Tactics Analysis Group, *Directory of Organizations and Activities Engaged or Interested in War Gaming*, Washington, D.C.: Report AD-403272, 1962.

Wilson, A., *The Bomb and the Computer*, New York: Delta, 1968.

Articles

Models and Simulations

Adams, R. H., and Jenkins, J. L., "Simulation of Air Operations with the Air-Battle Model," *Operations Research*, 8 (September–October 1960), 600–615.

"Advantages and Limitations of Computer Simulation in Decisionmaking"— Report to Congress GAO B-163074, 1973.

Antosiewicz, H. A., "Analytic Study of War Games," *Naval Research Logistic Research Quarterly*, 2, 3 (September 1955), 181–108.

Berkovitz, L. D., and Dresher, M., "A Game Theory Analysis of Tactical Air War," *Operations Research*, 7 (1959), 599–620.

Brotman, L., and Seid, B., "Digital Simulation of a Massed-Bomber, Manned-Interceptor Encounter," *Operations Research*, 8 (May–June 1960), 421–423.

175

Caywood, T. E., and Thomas, C. J., "Applications of Game Theory in Fighter Versus Bomber Combat," *Journal of the Operations Research Society of America, 3* (1955), 402–411.

Engel, J. H., "A Verification of Lanchester's Law," *Operations Research, 2* (1954), 163–171.

Dalkey, N. C., "Soluble Nuclear War Models," *Management Science, 11* 9 (1965), 783–791.

Dresher, M., "Some Military Applications of the Theory of Games," Santa Monica, Calif.: The Rand Corporation, P-1849 (December 1959).

Driggs, I., "A Monte Carlo Model of Lanchester's Square Law," *Operations Research (JORSA), 4,* 2 (April 1956), 148–151.

Geisler, M. A., "A First Experiment in Logistics System Simulations," *Naval Research Logistics Quarterly, 7,* 1 (March 1960), 21-44.

Haywood, O. G., Jr., "Military Decision and the Mathematical Theory of Games," *Air University Quarterly Review, 4* (Summer 1950), 17–30.

Haywood, O. G., Jr., "Military Decision and Game Theory," *Operations Research, 2* (November 1954), 365–385.

Kahn, H., "Use of Different Monte Carlo Sampling Techniques," in H. A. Meyer (Ed.), *Symposium on Monte Carlo Methods*, New York: Wiley, 1956, 146–190.

Kahn, H., and Mann, I., "Monte Carlo," Santa Monica: The Rand Corporation P-1165 (July 1957).

Nolan, J. E., Jr., "Tactics and the Theory of Games: The Theory of Games Applied to the Battle of Guadalcanal," *Army, 11* (August 1960), 77–81.

Singer, J. D., "War Games: Validity and Interpretation," *Army, 16* (April 1966), 64–68.

Taylor, J. L., "Development and Application of a Terminal Air Battle Model," *Operations Research, 7* (November–December), 783–796.

Thrall, R. M., "An Air War Game," *Research Review*, (December 1953), 9–14.

Zimmerman, R. E., "A Monte Carlo Method for Military Analysis," in *Operations Research for Management, Vol. II*, Baltimore: Johns Hopkins Press, 1956.

Gaming

Averch, H., and Lavin, M. M., "Simulation of Decisionmaking in Crises: Three Manual Gaming Experiments," Santa Monica: The Rand Corporation, RM-4202-PR, (August 1964).

Banister, A. W., "The Case for Cold War Gaming in the Military Service," *Air University Review, 18,* 5 (July–August 1967), 49–52.

Brooks. R. S., "How it Works—the Navy Electronic Warfare Simulator," *U.S. Naval Institute Proceedings*, (September 1959), 147–148.

Cockrill, J. R., "The Validity of War Game Analysis," Annapolis, Md.: *U. S. Naval Institute Proceedings*, (September 1961), 59–70.

Deems, P. S., "War Gaming and Exercises," *Air University Quarterly Review, 9* (Winter 1956–1957), 98–126.

deLeon, P., "Scenario Designs: An Overview," Santa Monica: The Rand Corporation, R-1218-ARPA (June 1972).

De Quoy, A. W., "Operational War Gaming," *Armor*, *72* (September–October 1963), 34–40.

Ellis, J. W., Jr., and Greene, T. E., "The Contextual Study: A Structured Approach to the Study of Political and Military Aspects of Limited War," *Operations Research, 8* (1960), 639–651.

Helmer, O., "Strategic Gaming," Santa Monica: The Rand Corporation, P-1902 (1960).

Kahn, H., "War Gaming," Santa Monica: The Rand Corporation, P-1167 (July 1957).

Maloney, E. S., "Modern War Gaming: State of the Art," *Marine Corps Gazette*, (November 1960), pp. 10–12.

McDonald, T. J., "JCS Politico-Military Desk Games," *War Gaming Symposium, Second, 1964, Proceedings*, Washington, D.C.: Washington Operations Research Council 1964, 63–74.

McHugh, F. J., "The Oldest Training Device," *Naval Training Bulletin* (Fall, 1964), 8–12.

McNichols, G. R., "An Objective Appraisal of War Gaming," Washington, D.C.: U.S. Air Force, Operations Analysis Group, (1970).

Newton, J. M., et al, "Submarine Tactical Games," *Naval Research Reviews*, (June 1959), 14–17.

Paxson, E. W., "War Gaming," Santa Monica: The Rand Corporation, RM-3489-PR, (February 1963).

Renshaw, J. R. and Heuston, A., "The Game Monopologs," Santa Monica: The Rand Corporation, RM-1917-1-PR, (1960).

Specht, R. D., "Gaming as a Technique of Analysis," (Abstract), *Journal of the Operations Research Society of America, 3*, 1, (1955), 120.

Specht, R. D., "War Games," Santa Monica: The Rand Corporation, P–1041, (March 1957).

Thomas, C. J., "Military Gaming," in R. L. Ackoff (Ed.), *Progress in Operations Research, 1*, 5, New York: Wiley, 1961, 421–464.

Thomas, C. J., and Deemer, W. L., "The Role of Operational Gaming in Operations Research," *Operations Research, 5* (February 1957), 1–27.

Thomas, C. J., and McNichols, G. R., "Why People Play Games—Report of a Survey," Paper presented at the meeting of the Operations Research Society, Denver, June 1969.

War Games for Battalion, Regiment and Division," *Military Review*, (March 1941), 48–51.

Weiner, M. G., "Gaming," E. S. Quade and W. I. Boucher (Eds.), *Systems Analysis and Policy Planning: Applications in Defense*, New York: American Elsevier, 1968, 265–278.

Weiner, M. G., "An Introduction to War Games," Santa Monica, Calif.: The Rand Corporation, P-1773 (August 1959).

Weiner, M. G., "War Gaming Methodology," Santa Monica: The Rand Corporation, RM-2413, (July 1959).

Young, J. P., "A Survey of Historical Developments in War Games," Operations Research Office Staff Paper 98 (August 1959). (Now Research Analysis Corporation).

COMMENTARY

Books

Featherstone, *Naval War Games*, and other books: These are simple, small, readable books for the amateur war gaming buff. They are quite clearly British in style. These books can easily be skipped by those who wish to "get on with" serious war gaming, but they are fun and worth knowing about.

Livermore, *The American Kriegspiel:* This is the first book published in the United States on the subject of war gaming. It contains a brief introduction of gaming prior to 1898 and throughout presents heavy emphasis on planning factors and their use in making calculations. It stresses that the factors should be based on empirical evidence.

Five types of game are noted:

1. the technical game representing an engagement in all its detail
2. the grand tactical game representing the general outline of an extensive battle
3. the strategic game involving movements of armies over an extensive area over a period of several days or months
4. the fortification game, representing siege operation
5. the naval game

Livermore stresses the tactical game with a two map version running with a barrier between the players so that they cannot see the disposition of each other's troops. This is precisely how the game of double blind chess (also known as Kriegspiel) is played. For one seriously interested in war gaming it is genuinely worth the effort to go back and look at this classic. Too frequently what happens with books of this type is that everyone copies everyone else's bibliographical references and no one goes back to the original.

McHugh, *Fundamentals of War Gaming:* This is a first-class clear introduction to the fundamentals of war games. Chapter 1 covers simulation, war games, models, war-game models, the role of the war game director, the control group, the players, the spectators, specification of purposes, types, scope, and level of games; their numbers of sides, amount of intelligence, methods of evaluation, and basic simulation techniques. The value of war games is considered and their limitations and relationship to game theory are also discussed. In Chapter 2 an excellent history of war games is presented. Chapter 3 deals with rules, procedures, and data. Chapter 4 is concerned with manual games. These games are discussed in somewhat more detail in a separate publication by McHugh. Chapter 5 is devoted to the Navy Electronics Warfare Simulator (NEWS) and the gaming based upon its use. Chapter 6 discusses computers and computer games. The weaknesses of this otherwise-excellent book lie in its omissions, not in its content. A critical discussion of the success or lack of success of the games at the Naval War College would have been an excellent addition. A summary of the lessons learned from the many years of gaming at the Naval War College is called for; it is a pity that it was not presented here.

Morschauser, *How to Play Games in Miniature:* This is another good book for the gaming buff. The games described are somewhat differentiated according to weapons characteristics. He suggests three classifications of games into "the shock period, the musket period, and the modern period." The book contains several useful appendices on magazines and books for war gamers and places from which to obtain equipment.

Richardson, *Arms and Insecurity:* This, together with its companion volume which supplies the statistics of deadly quarrels, presents some interesting hypotheses and models concerning the nature of escalation. The book represents an early attempt at mathematical building in the realm of diplomatic and military affairs. It is easy to observe with the benefit of hindsight that many of Richardson's mathematical models are grossly oversimplified and might strike some as mechanistic in the extreme. Nevertheless, they represent an important first step in applying scientific method and mathematical model-building to an extremely important subject.

Phillips, *The Roots of Strategy:* This collection of readings of basic works on strategy is noted especially because of its translation of the earliest military classic on the principles of war written by the Chinese general, Sun Tzu, in the 6th century, B.C. Had this treatise been written 2,000 years later it would not have been out-of-date.

Ulhaner (Ed.), *Psychological Research in National Defense Today:* It is extremely easy for different researchers with parallel interests to be almost oblivious of related work going on in somewhat different areas. The languages are different; the peer group is different; and the publications are not the same. Every now and then, however, it is important to cross the lines and to make sure that a political scientist, a military analyst, or a game theorist becomes aware of what the psychologist and those who study human factors have to offer to help modify their own thought. This book serves as a connection into that literature.

Articles

The articles listed in the bibliography are broken down into the two categories of (1) models and simulations and (2) gaming. The literature is considerable, and only a few items are discussed here.

Adams and Jenkins, "Simulations of Air Operations with the Air Battle Model": The air-battle model is a very large-scale simulation of a two-sided global war which had as its origin earlier theoretical work done at Rand. The model consists of a Plan Converter, the air-battle model proper, and the output programs. The air-battle model proper contains seven routines: (1) missile launching, (2) bomber launching, (3) tanker operation, (4) bomber cell handling, (5) attrition by enemy defenses, (6) target selection and reconnaissance, and (7) blast damage and radiation effects. It is an enormous model and it is extremely hard to find out whether in fact the size has been a blessing or a curse. I am innately skeptical of models of this size. However, it may be that tucked away in some classified document are the results that serve as a vindication for its usefulness. From the point of view of the interested gamer, the article is worth reading and it pays to ponder over questions, such as exactly who was meant to be using the results, what were the validation criteria to be supplied, and how well did it meet these criteria.

"Advantages and Limitations of Computer Simulation in Decision Making": This document is a report by the GAO to Congress. It is available at the cost of $1.00 (gratis to students and faculty) from the U.S. General Accounting Office, Room 6417, 441 G Street, N.W., Washington, D.C. 20548. It is a follow-up from a study entitled "Computer Simulations, War Gaming, and Contract Studies." The previous study covered appropriations of around $250,000,000 annually. This study is limited to data on 132 models and games throughout DOD, built at a cost of an average of around $280,000 each. The study is related to that of Shubik and Brewer on "Models, Simulations, and Games." Both are available and provide detailed statistics on the usage and value of models, simulations, and games.

Berkovitz and Dresher, "A Game Theory Analysis of Tactical Air War": This provides a dynamic game-theory model of a war viewed as a series of strikes or moves, consisting of counter-air, air defense,

180

and close-support operations for land forces. The mathematics is difficult and the model is simple in comparison with the actual phenomenon. The authors take into account factors such as the number of targets, accidents, antiaircraft fire, and replenishments. They construct a payoff function on the assumption that the major goal of air support is to assist ground forces in the capture of territory, and they use as a measure of success the distance of advance.

DeLeon, "Scenario Designs, An Overview": Although the construction of plausible, playable scenarios is a key item in the development of free-form military-diplomatic games and other military games, it is surprising that there is virtually no technical literature on the subject. There are a few isolated master scenario writers, such as H. DeWeerd. It appears that a background in history is a desirable prerequisite for scenario writing. This article by DeLeon provides one of the few sources for a well-thought-through discussion and description of the problems involved in the generating of a reasonable scenario.

Engel, "A Verification of Lanchester's Law": During World War I, Lanchester pointed out the importance of the concentration of troops in modern combat. He described the engagement in terms of a set of differential equations and concluded that the strength of a combat force is proportional to the square of the number of combatants entering the engagement. Since that time there have been many variants and modification of Lanchester's Law. This particular example provides a study of the validity of Lanchester-type equations applied to the combat casualties suffered during the battle of Iwo Jima.

Kahn, "War Gaming", Paxson, "War Gaming", and Weiner, "An Introduction to War Games": These are three excellent introductions to the subject of war gaming written by three of the top professionals. All three are well worth reading.

Thomas and Deemer, "The Role of Operational Gaming in Operations Research"; and Thomas and McNichols, "Why People Play Games": In 1957 Thomas and Deemer published a careful, thoughtful survey on the role of operational gaming in operations research. In 1969 Thomas and McNichols surveyed around 100 gaming specialists to present a balanced view of the disadvantages and virtues of the use of games. The new survey shows how the range of applications has broadened. Even though the first article was written in 1957, it still merits reading and contains a good general assessment of the methodological and practical strengths and weaknesses of gaming.

11 Gaming and Related Topics: A Literature Guide

i

This chapter is unashamedly eclectic. Some small bibliographies are given on subjects related to gaming, which, at least in the estimation of the author, are enjoyable and instructive reading.

GAMES AND GAMESMANSHIP

The books presented under the title of "Games and Gamesmanship" are a mixed bag which could be further classified into those which deal with the study and history of games, those which are strictly concerned with games surveys, those which stress analogies, and other books which deal with the psychological aspects of the players. Under the first category, the works of Avedon and Sutton-Smith, Bell, Falkener, and Murray all belong. Leggett, Schuh, Smith, and Takagawa describe or pose games. Cho-Yo and Boorman stress the analogies between games and war or revolutionary strategy. The remaining books deal with the psychological aspects of the players, although in the case of Yardley, the specific game feature of poker is also stressed.

Books

Avedon, E. M., and Sutton-Smith, B., *The Study of Games*, New York: Wiley, 1971.
Bell, R. C., *Board and Table Games*, Fairlawn, N.J.: Oxford University Press, 1960.
Berne, E., *Games People Play*, New York: Grove Press, 1964.
Boorman, S. A., *The Protracted Game: A Wei-Ch-i Interpretation of Maoist Revolutionary Strategy*, Fairlawn, N.J.: Oxford University Press, 1969.

182

Cho Yo, *Japanese Chess: The Science and Art of War or Struggle, Philosophically Treated*, Chicago: The Press Club of Chicago, 1905.

Falkener, E., *Games Ancient and Oriental and How to Play Them*, New York: Dover Publications, 1961.

Goffman, E., *The Presentation of Self in Everyday Life*, New York: Doubleday, 1959.

Leggett, T., *Shogi: Japan's Game of Strategy*, Englewood Cliffs, N.J.: Prentice-Hall.

Murray, H. J. R., *A History of Board Games other than Chess*, Oxford: Clarendon Press, 1952.

Opie, I. and Opie, P., *Children's Games in Street and Playground*, Fairlawn, N.J.: Oxford University Press, 1959.

Potter, S., *The Theory and Practice of Gamesmanship*, London: Rupert Hart-Davis, 1947; New York: Holt, 1948.

Schuh, F., *The Master Book of Mathematical Recreations*, New York: Dover, 1968.

Smith, A., *The Game of Go*, Rutland, Vt.: Tuttle, 1956.

Spolin, V., *Improvisation for the Theater*, Evanston, Ill.: Northwestern University Press, 1963.

Takagawa, K., *To Play Go*, Tokyo: Japanese Go Association, 1956.

Yardley, H. O., *The Education of a Poker Player*, New York: Simon & Schuster, 1957.

Commentary

Avedon and Sutton-Smith, *The Study of Games*: This is a highly worthwhile and encyclopedic compilation by two self-confessed game buffs. Each chapter contains readings and a bibliography. It is multidisciplinary in approach but like all multidisciplinary approaches, including my own, it is always less interdisciplinary than is desired by the reviewer when he considers it in light of his own discipline. After an introduction with an essay by P. G. Brewster on the importance of the collecting and study of games, several chapters cover the history and origin of games, viewing them from historical sources, anthropological sources, folklore sources, and other general sources. These chapters include such goodies as the first antigambling laws which appear to have been instituted in Rome about 2,000 years ago, and the banning of golf by the Scottish Parliament in 1457, as well as an essay on the games and sports that appear in the works of Shakespeare. Section II is on the usage of games and covers games for military uses, games in education, and diagnostic and treatment procedures. It does not discuss games for experimental purposes and the chapters both on military and industrial uses are rather light. The chapter on business and indus-

183

trial games is probably the weakest and slightest in the whole book. The chapter on games in education contains a useful bibliography but scarcely joins issues on the problems in the use of games in education. It does, however, contain a brief essay by I. Kraft entitled "Pedagogical Futility in Fun and Games?" This brief note is probably overcritical and overly polemical in style, yet it counteracts the "Oh, wow!" feeling that is sometimes evinced by gaming enthusiasts. The last section of the book deals with the structure and function of games. This includes three chapters on games in social science, games as structure, and the function of games. The chapter on games as structure is an interesting but somewhat unsatisfactory exercise in taxonomy. The classifications of Culin, Murray, Bell, Huizinga, Caillois, and several others are noted. A bow is made toward the theory of games, but little, if any indication is given concerning its contribution to the classification of games. Nevertheless, this is an interesting chapter in the questions that it raises and in the variety of classifications that it suggests.

Berne, *Games People Play*: Frequently the amateur's hero is anathema to the professional. I found this book stimulating, extremely easy to read, amusing, not particularly deep, but suggestive and interesting. However, not being a psychiatrist, I do not feel competent to judge the scholarly merit of this work. I was first led to suspect that it might in fact have some merit by the vehemence of the attack of some of Berne's professional colleagues. It sounded like a cross between sour grapes at not having cashed in on a good thing first and an anxiety reaction against having a member of the clergy preach too freely and in the vulgar tongue to the masses.

Cho Yo, *Japanese Chess, The Science and Art of War or Struggle, Philosophically Treated:* This is a thoroughly enjoyable collector's item and is a philosophical treatise written in 1905 by a Japanese scholar and admirer of the United States. It has been reviewed in detail in M. Shubik and G. D. Brewer, "Reviews of Selected Books and Articles on Gaming and Simulation," Santa Monica: The Rand Corporation, R-732-ARPA (June 1972). However, the book contains a classical warning to kibitzers which I feel merits being included here in full:

The author, when a mere boy, watching his grandfather playing *IGO* was told once a while by his mother that he should not disturb the welfare of the players; and she referred to the square pit on the back of the chess board and *IGO*-board. She stated that when bystanders would make trouble or lead rough conducts around players, or say or remark or suggest about plans or take the side of one, or when one player would have acted any mean unmanly unchivalrous campaign on the stage of struggles, the player himself so provoked could punish the impolite, unresponsible fellow by killing the offender on the spot and by putting his head chopped off on the back pit turned upside down. The mother said that it was for the purpose to have the hollow part, and that the killed deserved to have been punished because of a violation of strict, fundamental laws, and ethical rules of etiquette of the *Samraism*, the first principle of the then governing class of people. (p. 27)

Murray, *A History of Board Games Other than Chess*: This is a sequel written approximately forty years after the author's enormous and distinguished *History of Chess* (Oxford University Press, 1913). It covers the origin, locale and rules of play for over 217 board games in a manner characterized by first-class, painstaking scholarship.

Yardley, *The Education of a Poker Player*: This book is a delight. It is written by a professional poker player who was also a U.S. agent in China. Part I is entitled "Three Poker Stories," Part II is also entitled "Three Poker Stories," and Part III is entitled "Other Poker Games." A dispassionate discourse is provided on five-card draw, jacks or better, five-card stud, five-card draw, deuces-wild with the joker, five-card draw, low ball, seven-card stud, high-low, and many variations. Each game is illustrated with a story that develops both the aspects of the mathematical odds of playing poker and the psychology of the game. This includes a discussion of gambler's greed, twitching when you have a full house, cheating, displaying sweating palms, and all of the other aspects of poker that make it more than a game involving the pure mathematical odds and yet a game where good play is subservient to these odds. As Damon Runyon once observed, "The race is not always to the swiftest nor the battle to the strong, but that is the way to lay your dough."

EDUCATION

The few books suggested below provide some lead into the pioneering work in understanding learning as evinced by Piaget and Brunner, and books for those interested in using games in the classroom today. The chapter on games for education by Avedon and Sutton-Smith and the listing of games, together with their description, by Tansey and Unwin, and by Twelker, provide more than enough references for anyone who wishes to seriously pursue this use of gaming.

Books

Avedon, E. M., and Sutton-Smith, B., *The Study of Games*, New York: Wiley, 1971.

Boocock, S. J., and Schild, F. O., *Simulation Games in Learning*, Beverly Hills: Sage, 1968.

Brunner, R. D., *Towards a Theory of Instruction*, Cambridge, Mass.: Harvard University Press, 1969.

Dalton, R., et al., *Simulation Games in Geography*, New York: Macmillan 1972.

Piaget, J., *Play, Dreams, and Imitation in Childhood*, London: Routledge & Kegan Paul, 1962 (originally published in 1951).

Reilly, M. (Ed.), *Play as Exploratory Learning*, Beverly Hills: Sage, 1974.

Tansey, P. J., and Unwin, D. J., *Simulation and Gaming in Education*, London: Methuen, 1969.

Twelker, P. A., *Instructional Simulation Systems*, Corvallis, Ore.: Continuing Education Publications, 1969.

Wing, R. L. et al., *The Production and Evaluation of Three Computer-Based Economics Games for the Sixth Grade*, Westchester County, N.Y.: Board of Cooperative Educational Services, 1967.

URBAN GAMING

Only three references are noted for urban gaming. They, however, should provide an adequate introduction in the sense that Duke and Feldt provide descriptions of Metropolis and the Cornell Land Use Game, and these two are among the largest and most widely used games for this type of gaming. The third book by Brewer is addressed primarily toward problems in the construction of large-scale simulations. It is included here not as an implicit criticism of the uses of Metropolis or CLUG, but as a reminder and a warning to those who wish to go from relatively simple games for teaching purposes to large-scale projects and simulations for operational purposes that the change in the problems to be faced is not merely that of size, but fundamental new difficulties in planning and organization appear. It is a long way from the game for teaching or experimentation to the operational simulation that is of any use.

Books

Brewer, G., *Politicians, Bureaucrats, and the Consultant*, New York: Basic Books, 1973.

Duke, R., *Metropolis: The Urban Systems Game*, Beverly Hills: Sage, 1974.

Feldt, A. G., *The Community Land Use Game*, Ithaca, N.Y.: Cornell University, Division of Urban Studies of the Center for Housing and Environmental Studies, 1968.

Commentary

Brewer, *Politicians, Bureaucrats and the Consultant*: This book, whose subtitle is "A Critique of Urban Problem Solving," presents a carefully documented examination of two large-scale simulations applied to urban problems. The relationship among the interested parties to the building of these simulations is examined in detail. The Brewer analysis and discussion is a must for anyone involved in the construction and utilization of a large game or simulation in which there are several different interested parties involved and in which it is quite likely that the builders, users, and sponsors may all be different groups with different purposes to be served.

Feldt, A. G., *The Community Land Use Game (CLUG)*: This book provides a manual version of CLUG capable of being handled with two operators and three to eighteen players. It is noted that for any more complicated version a computer is necessary. In this game, each of three teams starts with $100,000 of capital with which they seek to buy land, construct commercial or residential properties, and make a profit through buying and selling of land or through residential, industrial, and service establishments. The initial structure of the game provides (1) the location and efficiency of a major highway network; (2) the location of the utility plant providing municipal services; (3) the location of the major points of access with the outside world; (4) the system of real estate taxation to support building and maintenance of community services; and (5) the range and type of municipal land uses. Money enters and leaves the community from the outside world through payment for goods and services. It requires careful planning and management by the players to control transportation costs, municipal services, and renovation costs as a function of how and where they build. The end of the game is purposely not clearly defined, nor is winning, although return on investment does provide a criterion. After about 10 hours of play, which covers between 15 and 20 rounds, most players are well-acquainted with the operation of the game and its potentialities. Among the activities in the game are the assessment of real property, the payment of taxes, the setting of the tax rate, providing for utilities, construction, demolition, or movement of buildings, renovation, and many others. Zoning rules and changes have also been considered. There is also a variant of the game to permit agricultural land-use modification to study land-changes from use in agriculture to urban usage.

OTHER TOPICS INCLUDING
SOCIOLOGY, PSYCHOLOGY AND
ARTIFICIAL INTELLIGENCE

This "catchall" bibliography includes the two important books on games by Caillois and Huizinga, both of which are reviewed below, as well as two fine books at a very high level of exposition on human communication by Cherry and Singh. There are several references to books on artificial intelligence. These are somewhat out-of-date, as a perusal of the recent work of Minsky and others would show. Nevertheless, especially *Computers and Thought*, by Feigenbaum and Feldman, is highly readable and provides several insights into the fascinating problems of how to construct artificial players. These are certainly of direct interest to the experimental gamer.

Books

Ashby, W. R., *Design for a Brain*, New York: Wiley, 1960.
Bergler, E., *The Psychology of Gambling*, New York: Hill & Wang, 1957.
Caillois, R., *Man Play and Games*, London: Thomas and Hudson, Ltd., 1962.
Cherry, C., *On Human Communication*, Cambridge, Mass.: M.I.T. Press, 1967.
Feigenbaum, E., and Feldman, J., *Computers and Thought*, New York: McGraw-Hill, 1963.
Guetzkow, H., and Kotler, P. (Ed.), *Simulation in Social and Administrative Science*, Englewood Cliffs, N.J.: Prentice-Hall, 1970.
Huizinga, J., *Homo Ludens* (Translated by R. F. D. Hull, London, 1949), Boston: Beacon Press, 1950.
Sass, M. A., and Wilkinson, W. D., *Computer Augmentation of Human Reasoning*, Washington, D.C.: Spartan Books, 1965.
Singh, J., *Great Ideas in Information Theory: Language and Cybernetics*, New York: Dover, 1966.
Von Neumann, J., *The Computer and the Brain*, New Haven: Yale University Press, 1958.

Commentary

Caillois, *Of Man, Play and Games*: In this excellent book, Caillois pays tribute to the earlier work of Huizinga but believes that he minimized the description of the diverse form of play and the many needs that play serves in various cultures. Caillois defines play as free, separate, uncertain, and unproductive, yet regulated and make-believe. The components of play are categorized under four major categories: (1) Agon (competition); (2) Alea (chance); (3) Mimicry (simulation); and (4) Ilinx (vertigo). He notes that under certain conditions the categories may be linked. For example, many Australian, American Indian, and African cultures illustrate the Mimicry-Ilinx complex in their emphasis on masks and states of possession. Ancient China and Rome, on the other hand, reflect the opposing principle, Agon-Alea, in stressing order, hierarchy, codification, and other devices of the interaction between competitive merit and accident, or chance of birth. The book is divided into two parts, the first entitled "Play and Games: Theme," the second "Play and Games: Variation." In the first part there are five chapters covering (1) the definition of play; (2) the classification of games; (3) the social function of games; (4) the corruption of games; and (5) toward a sociology derived from games. The second part offers (6) an expanded theory of games, (7) a discussion of simulation and vertigo, (8) a discussion of competition and chance; and (9) the last chapter, entitled "Revivals in the Modern World," deals with the modern role of Agon, Alea, Mimicry, and Ilinx. One minor criticism of Caillois is that he shows a fundamental misunderstanding of the role of economics when he notes that in some games property is exchanged, but no goods are produced, "Hence, play is an occasion of pure waste" (p. 5). He does not appear to understand that exchange alone may create value. He stresses one important criterion of being in a game is the possibility of being able to leave, by saying, "I am not playing any more." Games can be ruined by the nihilist or spoilsport who denounces the rules as absurd and conventional, and who refuses to play because the game is meaningless. Caillois' stress on the relationship between the context of the game world and the outside world is of considerable importance to all who build formal models. A major theme developed by the author is his contrast of societies based on Mimicry and Ilinx which he terms "Dionysian" societies, and societies based on Agon and Alea which he considers orderly or "rational" societies. He observes that primitive societies are ruled equally by masks and possession. Conversely, "rational" societies are ruled by offices, careers, codes, and fixed and hierarchical privileges, and merit and hereditary positions seem to complement each other. This book is virtually a "must" for the educated gamer.

Huizinga, *Homo Ludens, A Study of the Play Element in Culture*: This book, written in 1938 by an important historian of culture, is regarded by many as a classic. It is an exposition of the role of the instinct of play in law, poetry, military affairs, philosophy, art, and other aspects of civilization. The author provides a deep linguistic study of the nomenclatures of games, competition, and play in many languages. The nature and significance of play as a cultural phenomenon is considered, and the role of play in games in Western civilization is analyzed. Unfortunately, the work appears to suffer from a lack of interest or understanding in science and nonliterary approaches to scholarship. For example, the discussion of winning is hampered by Huizinga's lack of ability to distinguish clearly between zero-sum and nonconstant sum games. Comments such as his remarks on the stock exchange (p. 52) show that he basically does not understand the mechanism he is describing. This is a valuable and scholarly book and the reader who has both a healthy skepticism and a good scientific background can benefit from the insight he will gain reading it, while at the same time he can avoid the pitfall into which the author himself has fallen. The study of games and their role in society is too important a subject to be investigated only by extended *belles lettres*, essays, and other literary approaches alone.

Singh, *Great Ideas in Information Theory, Language and Cybernetics*: This is a first-class, reasonably priced paperback presenting a highly readable, yet technically sound introduction to information theory, problems in coding, transmission and redundancy, neuronetworks, Turing machines, and other aspects of artificial intelligence, including game-playing machines. This book serves as a connection at an introductory level between the interests of those concerned with gaming and those concerned with artificial intelligence. It is science writing at its best.

LITERATURE

Three novels are noted and reviewed below. Each of them is directly relevant to some aspect of gaming. There is also an opera entitled "Gambler's Luck." Three other novels also come to mind, although they are not quite so directly relevant to gaming: They each present an extremely interesting study of personal decision making styles. They are: *Oblomov*, by Goncharov; *The Trial*, by Kafka; and *The Good Soldier Schweik*, by Capek.

Books

Brunner, J., *The Squares of the City*, New York: Ballantine Books, 1965.
Dostoevsky, F., "The Gambler," in *The Gambler, Bobok, A Nasty Story*, Baltimore: Penguin, 1966.
Hesse, H., *Magister Ludi*, New York: Holt, Rinehart and Winston, 1969.

190

Commentary

Brunner, *The Squares of the City*: This is a thoroughly enjoyable and worthwhile science-fiction story where the whole novel is designed, move-for-move, around a world championship chess game played by William Steinitz and Mikhail Tchigorin in 1892 in Havana. The chief character is an Australian-born traffic analyst who had been called to an imaginary South American country to redesign the traffic flow for the capital city. As the plot unfolds, he discovers that he has been selected because he appears to be an apolitical technocrat. The actual purpose for his employment was to help the dictator and his minister of communications to reduce some trouble spots in the city, giving the appearance of doing it not for political purposes, but through the dictates of objective technocratic progress. The contrast between the richness of political context and the abstraction of the chess game makes this novel not merely enjoyable reading but valuable reading for those seriously interested in the arcane arts of model-building and gaming.

Dostoevsky, "The Gambler": This is a tense, chilling, brilliant description of the addicted gambler. I find it somewhat surprising that the scholarly literature on gambling is as slight as it appears to be. In the previous bibliography, one book, that by Bergler, is noted. In spite of the considerable interest in decision-making under risk, the connection between the study of gambling, risk-taking and theories of risk-taking is surprisingly slight. This alone makes the fictional writing about gambling of considerable worth to the gamer.

There is an article on gambling by Devereux in the *International Encyclopedia of the Social Sciences*. The bibliography attached to this article is relatively slight, and the author also notes that the available studies of work on gambling are surprisingly few.

Hermann Hesse, *Magister Ludi (The Glass Bead Game)*: Although it might be regarded by some as stretching the point to include this novel in a book on gaming, I believe that the development of the theme of the highly abstract gaming model as both an all-encompassing intellectual exercise and at the same time a dangerously sterile procedure is of considerable interest to the sensitive modeler of abstract representations of human affairs. Several quotations may help to illustrate this point:

> These rules, the sign language and grammar of the Game, constitute a kind of highly developed secret language drawing upon several sciences and arts, but especially mathematics and music (and/or musicology), and capable of expressing and establishing interrelationships between the content and con-

191

clusions of nearly all scholarly disciplines. The Glass Bead Game is thus a mode of playing with the total content and values of our culture; it plays with them as, say, a painter might have played with the colors on his palette. (p. 6)

In the novel, the main character visits a Benedictine historian who is not enamored with the Glass Bead Game:

You mathematicians and Glass Bead Game players have distilled a kind of world history to suit your own tastes. It consists of nothing but the history of ideas and of art. Your history is bloodless and lacking in reality. You know all about the decay of Latin syntax in the second or third centuries and don't know a thing about Alexander or Caesar or Jesus Christ. You treat world history as a mathematician does mathematics, in which nothing but laws and formulas exist, no reality, no good and evil, no time, no yesterday, no tomorrow, nothing but an eternal, shallow mathematical present. (pp. 150–151)

In the applications of gaming, be they to operational problems involving economics, war, or societal affairs, or to teaching, the conflict between mathematical abstraction and simplicity on the one hand and societal richness and historical content on the other hand is always present. Neither a pure literary nor a wholly mathematical approach to scholarship holds all the answers. When viewed in this light, this novel has something to teach even the most hardened military engineer anxious to "get on with the war game"—if he is still willing to learn.

Author Index

Abt, C. C., 30
Adams, R. H., 180
Alger, C. F., 131
Alker, H., 167
Allen, Layman E., 30, 126
Avedon, E. M., 44, 133, 182, 183

Balderston, F. E., 140
Barringer, R., 173
Barton, R. E., 53, 55, 132, 138
Bastable, C. W., 141
Bell, R. C., 133, 182
Bergler, E., 43
Berne, E., 184
Bixenstine, V. E., 158
Bloomfield, L., 167
Bonoma, T. V., 157
Boocock, S. S., 30
Boorman, S. H., 171, 182
Boucher, W. I., 173
Bowen, K. G., 48
Bowley, A. L., 154, 157
Brewer, G. D., 122, 131, 132, 167, 180
Brewster, P. G., 183
Brody, R. A., 131, 172
Brunner, R. N., 167

Caillois, R., 40, 44, 184
Caplow, T., 160
Chamberlin, E. H., 159
Chambers, N., 158
Chammah, A. M., 154
Chertkoff, J. M., 160, 161
Cho-Yo, 182, 184
Churchill, N. C., 139

Cohen, K. J., 138
Coleman, J., 30
Couger, J. R., 135
Cournot, A. A., 154, 157
Culin, S., 184

Dalkey, N. C., 36
Davis, M. D., 130
Deemer, W. L., 181
Deep, S. D., 141
DeLeon, P., 181
Deutsch, M., 167
Dostoevsky, F., 43
Dresher, M., 131, 180

Eisenberg, H., 162
Engel, J. H., 181

Falkener, E., 182
Featherstone, D. F., 178
Feigenbaum, E. A., 133
Feldman, J., 133
Fellner, W., 157
Finger, J. M., 115
Fishman, G. S., 114, 115
Forrester, J. W., 140
Fouraker, L. E., 153, 157
Friedman, J., 160

Gamson, W. H., 160
Geisler, M. A., 141
Gibbs, G. I., 128
Goeldner, C. R., 142
Goffman, E., 43
Goldhamer, H., 167
Gordon, D. J., 130, 155

193

Graham, R. G., 128, 139, 143
Graves, R. L., 143
Gray, C. F., 128, 139, 143
Greenlaw, H., 129
Guetzkow, H., 74, 80, 131, 171, 172
Guyer, M., 129, 130, 145, 155

Haldi, J., 139
Harnett, D. L., 153
Helmer, O., 36
Henshaw, R. C., 139
Herron, L. W., 129
Hillel, Bar, 39
Hoggatt, A., 38, 88, 140, 160, 161
Horn, L., 128
Huizinga, J., 40, 184

Ikle, F., 172

Jackson, J. R., 139
Jenkins, J. L., 180
Jones, B., 148

Kahn, H., 181
Kennedy, John, 67
Kaplan, R. J., 167
Kidder, S. J., 127
Kiviat, P. J., 53, 56, 114, 115
Kraft, I., 184
Krasnow, H. S., 53

Laponce, J. A., 132
Levine, R. A., 174
Levitan, R. L., 73
Lieberman, B., 161
Livermore, W. R., 178
Lockhart, S., 37

Malcolm, D. S., 129
Maschler, M., 161
Mayberry, J. P., 115
McHugh, F. J., 131, 166, 178, 179
McKenney, J., 141
McNichols, G. R., 181
McRaith, J. F., 141
Meeker, R. J., 88
Meier, R. C., 129
Merikallio, R. A., 53, 56
Mickolus, E., 132
Minsky, M., 39, 133
Morchauser, J., 179

Morehous, L. G., 161
Murray, H. J. R., 133, 182, 184

Nash, J., 155, 157
Naylor, T. H., 115, 129, 130, 132, 140
Newell, W. T., 129
Noel, R. C., 89, 90, 131, 172

Orcutt, G. H., 140

Papert, S., 39
Paxson, E. W., 181
Pazer, H. L., 129
Perkel, B., 129
Phillips, J. L., 179
Piaget, J., 32
Pool, I. de Sola, 167

Quade, E. S., 173

Rapoport, Anatol, 97, 130, 154, 155, 173
Rawdon, R. H., 129
Richardson, L. F., 179
Riley, V., 130

Savage, E., 132
Schelling, T. C., 156, 167, 173, 174
Schild, E. O., 30
Schlenker, B. R., 157
Schrieber, A. N., 129, 140
Schuh, F., 182
Selten, R., 161
Shubik, M., 38, 73, 97, 130, 132, 142, 153, 154, 156, 160, 162, 163, 166, 180
Shure, G., 88
Siegel, A. I., 153, 157
Smith, A., 182
Smith, V., 160, 164
Smoker, P., 132
Snyder, R., 132, 167, 172
Sun Tzu, 179
Sutton-Smith, B., 44, 133, 182, 183

Takagawa, K., 182
Tedeschi, J. T., 157
Thomas, C. J., 181
Thorelli, H. B., 143
Tocher, K. D., 53
Twelker, P. A., 127

Ulhaner, H. E., 179

Van Horn, R. L., 115, 116

Wagner, H., 130
Weiner, M. G., 181
Whaley, B., 173
Wilson, A., 131, 158, 166, 173
Wing, R. L., 30

Wohlstetter, A., 167, 174
Wolf, G., 38, 97, 162

Yardley, H. O., 41, 182
Young, J. P., 130

Zeuthen, F., 157
Zuckerman, D., 128

195

Subject Index

Abstraction, 7, 71
Abt Associates, 126
Academic Games Associates, 126
Accounting, 143
Accounting Education, 139
Accounting System, 70
Acoustics, Tile, 95
Activities, 56
Administration, 77
Administrative Science Quarterly,
    125
Adult education, 27
Advertising, 31, 162
    joint maximum, 163
Advocacy, 36
Aerospace, 135
Agent, 13
Aggregation, 7, 105
AIMS, 142
Air battles, 180
Air defense, 180
Air Force logistics, 94
Air University Review, 125
Allocation games, 102
Alternatives, 23
Amateur gaming, 27, 125, 128
American Behavioral Scientist, 125,
    128
American Management Association
    (AMA), 85, 143
AMA Business Game, 36
Analog
    computer, 8
    equipment, 63

Analysis, 81, 115
    comparative, 67
    equal-share, 161
    factor, 114, 133
    force-structure, 167
    military, 179
    sensitivity, 79, 107, 108
    spectral, 114
Analysis problems, 115
Analysis systems, 71
Analysis variance, 114
Anti-aircraft fire, 181
Anticipating, 49
Anti-gambling laws, 183
Anti-games, 45
Arbitrager, 41
Army's Strategy and Tactics Analysis
    Group, 126
Artifacts, 11
Artificial intelligence, 32, 38
Artificial player, 38, 39, 40
Aspiration levels, 157
Assassinations, 43
Assignment of costs, 62
Audio systems, 87
Audio tapes, 4
Automation in classroom, 78
Avalon Hill Company, 126

Babcock, Allen Co., 84, 90
Banking, 135
Bargaining, 11, 31, 39, 158
    set, 158, 163

Baseball, 43
Basketball, 43
Batch, 84
Battle area, 165
Battle exercises, 33
Battlefield, 6
Beat-the-average, 20, 163
Behavior
    continuity of, 107
    bargaining, 158
    cooperative, 162
    cyclical, 163
    limiting, 105, 107
    monopolistic, 162
    pathological risk, 43
    rational, 22
    small group, 86
    socially rational, 32
    two-person, 158
    voting, 166
    within a structure of rules, 45
Behavior systems, 113
Behavioral science, 8, 18, 106, 168, 173
Behavioral Science, 125
Behavioral Science Laboratory, 87
Behavioral Science Research
    Laboratory, 126
Behavioral theory, 24
Benefits, 68, 69
Berkeley, 78, 87
    Center for Research in
        Management Science at, 83
Bibliography, 122, 126, 128, 134, 184
Bidding, 125, 154
Binary mode, 55
"Black box phenomenon," 113
Blackmail, 16
Board games, 8, 76
BOCES (Board of Cooperational
    Educational Services), 85
Body movement, 22
Bomber launching, 180
Booby trap, 114
Boredom, 23, 112
Bowley point, 154
Box office receipts, 101
Brainstorming, 35
Breadboard, 89

Bridge, 7, 22, 44
Briefing, 80, 111
Brookings Institute, 8
Builders, 29, 34, 122
Building, 62
Buildup, 65
Bureaucracy
    decision-making, 31
    life, 172
    organization, 27
    routines, 33
    structure, 24
Business games, 6, 12, 22, 26, 27, 37, 39, 72, 128, 183
Business management, 134
Business schools, 135
Business simulation, 129
Byefleet, 126

Calculated probabilities, 41
Calculated risk, 41, 42
Canadian Army Journal, 125
Captive, 56
Card games, 7
Caretaker, 101
Carnegie Mellon University, 84
Carnegie Tech Game, 30, 31, 143,
Catalogue of War Gaming Models, 131
Categorized, 127
Center for Research in Management
    Science, 83
Central city decay, 36
Characteristic function, 15, 163
Charades, 44
Checking, 110
Chess, 6, 7, 12, 17, 20, 39, 44
Children, disadvantaged, 30
China, 171
Choreography, 74, 111
Circuses, 41
Classification, 126
    of gaming literature, 119
Classroom, 69
    automation in, 78
Clearinghouse, 86
Clocks, game, 93
Coalition, 15, 16

cooperative, 20
Coalition games, 161
Coalitional form, 11
   three-person, 160
COBOL, 56
Coding, 11, 22
College, war, 85
Colonel, 23
Combinatorics, 123
Communication, 31, 34, 54, 97,
   144, 159
   channel of, 19
   limited, 19
   structure, 158
Community action, 31
Community Response, 85
Competition, 38
   price of, 163
Computation costs, 77
Computation damage, 93
Computation errors, 77
Computer-based laboratory at SDC,
   83
Computers, 61, 73, 79, 83, 84
   costs of, 59
   high-speed digital, 8, 35
   instruction for use of, 39, 85
   machines, 54
   in simulation, 138
Computer center, 86
Computer languages, 53, 54, 73, 77,
   PDP8, 88
   PDP5, 88
Computer systems, 56, 78
Concepts, solution, 11
Conflict, 115
   resolution, 125, 156
Conflict Resolution, Journal of, 124
Conflict theory, 158
Con-games, 43
Consistency, 50
Constraints, 102
Contingency prediction, 35
Continuity, 105,107
Control, 26, 38, 49, 50, 108, 116
   management, 76
   team, 81
Control group, 178
Controller, 5, 29, 34
Cooperation, *see* games

Cooperative behavior, 162
Cooperative solutions, 17, 18
Core, 21, 163, 164
Coronations, 41
Corporation, private, 27
Costs, 51,59, 68, 70, 77, 97, 103,
   111, 120
   administrative, 64
   allocation of, 68
   assignment of, 62
   buildup of, 65
   computation of, 77
   effectiveness of, 60
   investment, 62
   joint, 62, 68, 69, 75
   long-run, 64
   maintenance, 63
   of gaming, 4
   opportunity, 68, 69
   overhead, 69
   player, 64
   replacement, 63
   research and development, 62
   unallocated, 68
Crossword puzzles, 44, 101
Cues, 111
Curriculum games, 93
Customer, 115
Cyclical behavior, 163
Czech Experimental Theater, 31

Damage, computation of, 93
Data
   reduction of, 79
   sources of, 113
   time-series, 50
Data bank, 49, 62, 65, 77, 78, 82
Data base, 50
Data gathering, 49, 51, 67
Data organizing, 81
Data processing, 11, 22
Dean, 120
Death, 42, 46
Debriefing, 67, 68, 81, 111
Debugging, 61, 73
Decision making, 9, 31, 36, 39, 172
   conscious, 9
   goal-oriented, 9
   individual, 24
   level, 143

Decision-making (*cont'd*)
  military, 173
  multiperson, 9
Decision-making processes, 71
Decision Sciences, 125
Defense
  air, 180
  department of, 131
Defense planning, 173
Defense system, 73
Delphi, 35, 36
Democracy, 32, 85
Dependability, of data sources, 113
Design, 5
  game, 38
  laboratory, 87
  scenario, 181
  systems, 94
Destroyer, 109
Development, 61, 62
Device, data organizing, 81
Digital computer, 8
Dimension checking, 50, 105
Diplomacy, 29
Direction, 105
Director of war games, 178
Disadvantaged children, 30
Disaggregation, 106
Display, 78
Distribution, 76
Documentation, 67, 72, 76, 86, 104, 110
Dollar Auction Game, 45
Double acrostics, 57
Dress rehearsal, 33, 74
Dreyfus, H., 39
Duopoly, 154
Dynamics, solutions, 17
  game theory, 18
DYNAMO, 56

Econometrics, 37, 49
Economics, 6, 22, 94, 144, 157
  experimental, 146, 156
Education, 146
  adult, 27
  in games, 30, 31
Efficiency, 115
Egalitarian solution, 164
Elections, 166
Elementary games, 26

Engineering, 56, 168
Entertainment, 28
  games for, 76
Entertainment fee, 42
Environment-poor games, 10, 11, 22
Environment-rich games, 10, 11, 22
Equal-share analysis, 161
Equations, differential, 52
Equilibrium point, 19, 20
  competitive, 158, 162, 163, 164
  noncooperative, 19, 162
Equipment, 61
  analog, 63
Escrow, 156
Estimation of equations, 50
Ethicality measure, 159
Executions, public, 41
Executive Game, 139
Exercises
  diplomatic, 79
  gaming, 89, 100
  logistics of, 85, 98
  political-military, 102, 109, 112, 167, 174
Experimentation, 28, 53, 75, 144
  education, 146
  in gaming, 37, 38, 60, 70, 77, 78, 124, 129
  pilot, 40
  psychology of, 37, 112
Extensive form, 11
External symmetry, 25
Extremist groups, 43
Evaluation, 127
  doctrinal, 167
  of gaming literature, 119
  methods of, 178
  technical, 167
  weapons, 26, 58, 85
Evaluation error, 77
Evaluation operations, 91
Events, 56

Facial expression, 20
Facilities, laboratory, 61, 69
FAME Game, 73, 142
Feasibility, 29
  economic, 97
  technological, 97

Feedback, 53, 139
Field exercises, 33
  maneuvers in, 32
Fighter pilot, 42
Finance, 135
Fleet games, 93
  maneuvers, 40
Flexibility, space, 88
Floor space, 87
Flow, traffic, 85
Flowcharts, 75
Folk games, 32
Folklore, 183
Footnotes, 120
Force
  posture, 102
  structure, 86, 167
Forecasting, 36, 49
FORTRAN IV, 54, 56
Four-person games, 161
Free-form games, 9, 73, 74
Function
  characteristic, 163
  objective, 52
  payoff, 181
Functional form, 26
Functional game, 184

Gallery, observation, 95
Gambling, 6, 41, 42
Game
  clock, 93
  construction, 61, 71, 72
  history, 182
  structure of, 184
Game design, 38, 61
Game function, 184
Game room, 37
Game tree, 12
Games for Society, Business and War,
  3, 11
Game theory, 3, 9, 18, 22, 24, 25,
  106, 115, 161, 168, 173, 179, 184
  dynamic, 18, 155
Games, 124
  allocation, 102
  AMA Business Game, 36
  board, 8, 76
  business, 6, 12, 22, 26, 27, 37,
    39, 72, 183

card, 7
coalitional, 11, 20, 144, 158
community action, 31
constant-sum, 146
cooperative, 20, 155, 163, 165
curriculum, 93
diplomatic, 27, 34, 181
dynamic, 162
economics, 6
educational, 30, 31
elementary, 26
entertainment, 3, 6, 28, 76
environment-poor, 10, 11, 22
environment-rich, 10, 11, 22
Executive Game, 139
experimental, 28, 37, 38, 39, 60,
  70, 77, 78, 100, 124, 129, 144
extensive form, 11
fleet, 93
folk, 32
fortification, 178
four-person, 161
free-form, 9, 45, 74
functional, 135
general-purpose, 135
instructional, 126
international relations, 9
Japanese war, 37
land use, 6
large, 72, 75
man-machine, 86, 100
manual, 91, 179
matrix, 13, 37, 38, 45, 102, 145,
  155, 156, 161, 162, 167, 173
military, 6, 23, 24, 27, 34, 145,
  158, 166, 167, 168, 181
naval, 178
nonconstant sum, 18, 29, 31, 39
normal, 11
n-person, 24, 32
of skill, 51
of social status, 158
one-map, 93
one-shot, 155
operational, 4, 26, 28, 33, 35,
  38, 65, 71, 72, 74, 75, 80, 81
parlor, 6
planning, 35
political, 24, 125, 130, 166
political-military, 173
posture, 102

Games (*cont'd*)
  recreational, 6
  rigid-rule, 10, 45, 79
  sequential, 162
  solitary, 44
  strategic, 166, 178
  strategic form, 4
  tactical, 178
  technical, 178
  3 × 3, 161
  three-person, 163
  2 × 2, 37, 38, 45, 97, 155, 156, 162, 167, 173
  two-person, 18, 29, 39, 146, 165
  two-sided, 93
  war, 23, 27, 37, 70, 86, 91, 98, 126, 131, 178, 179
  World Game, 37
  zero-sum, 18, 29, 39, 146, 165
Gaming, 3, 18, 35, 107, 124, 126
  costs of, 4
  in diagnosis, 28
  in operations research, 3
  in teaching, 3, 27, 28, 29, 32, 84, 85
  in therapy, 3, 28
  in training, 3, 28, 32, 33, 85
  methodology of, 3
Gaming activities, 6
Gaming exercises, 89, 100
Gaming facilities, 4, 98
Gaming laboratories, 4, 98
Gaming literature, 119
Gaming manuals, 75
Gaming and Simulation College of the Institute for Management Sciences, 125, 126
Gamesmanship, 182
General
  Army, 42
  4-star, 4
The General, 125
General Motors, 7
Go, 171
Goals, 102, 103
Golf, 44, 183
Good Housekeeping, 74
Government, 27
GPSS, 56
Group formation, 11

Group therapy, 45
Guinea pigs, 37

Handbook, 71
Hardware, 54, 55
  military, 102
Harvard Business School, 31
Heuristics, 15
Hexagonal paving, 91
High School, 85
High-speed digital computers, 8, 35
High-speed memory, 57
History game, 183
Hockey, 43
Horse races, 41
Housekeeping, 110
Human factors, 38, 126, 133

IBM research labs, 78
IBM, 360 Model, 75, 89
IGO, 184
Income tax, 70
  laws, 73
Industrial College of the Armed Forces, 85
Influence, social, 158
Information, 11, 154
  incomplete, 154
  processing of, 23, 24, 39
  state of, 9
Information conditions, 12
Information set, 12
In-house, 100
Initiation ceremonies, 32
Input, 26, 54
Input-output, 75, 78
Institute for Defense Analysis, 125
Instruction
  computer-aided, 38, 85
  games for, 126
Insurance buying, 42
Integer programming, 32
Intelligence, 178
  artificial, 32, 38
  individual, 39
  social, 39
Intention, 26, 46, 101, 113, 116
International Federation of War Games, 126
International Journal of Game

Theory, 124
INS International Simulation, 80, 85, 109, 115, 167
Internation Simulation, 131
International problems, 125
International relations, 9, 50, 89, 166, 167, 172
INTOP, 30
Interpersonal interaction, 84
Inventory production, 8
Inventory scheduling, 32
Investment costs, 62
IPL, 56
Iwo Jima, 181

Japanese war gaming, 37, 171
Jigsaw puzzles, 44
Job shop, 109
Joint Chiefs of Staff Studies Analysis and Gaming Agency, 126
Joint War Games Agency, 131
Journal of Abnormal and Social Psychology, 125
Journal of Experimental and Social Psychology, 125
Journal of Experimental Psychology, 125
Journal of the Operations Research Society of America, 125
Journal of Personality and Social Psychology, 125

Kernal, 163
Keypunching, 85

Laboratory, 83, 86, 88, 95, 97, 141
    Berkeley, 83, 87
    commercial, 86
    design of, 61, 87
    discipline in, 111, 120
    facilities in, 61, 69, 145
    gaming, 86, 98
    language, 89
    Leuven, 83, 87
    NEWS, 83
    POLIS, 83, 88, 89
    Purdue, 83, 87
    Rand Logistics Simulation, 86
    SDC, 83

Language
    external machine, 55
    general purpose user, 55
    internal machine, 55
    laboratory, 89
    machine, 54, 55
    programming, 54, 78
    user-written, 55, 56

Las Vegas, 42
Launching
    missile, 180
    banker, 180
Lawyers, 23
Learning, 25, 29
Learning processes, 32
Learning Games Associates, 126
Leuven Laboratory for Experimental Social Psychology, 83, 87
Library, 61, 62, 119
License, 33
Life Career, 85
Limited communication, 19
Listing
    process, 52, 56
    program, 75
Literature
    classification of, 119
    search for, 119
    Soviet, 126
Logical switches, 52
Logistics, 85
    Air Force, 94
Logistics exercises, 97
Logistics Research Lab, 85
Logistics specialists, 94
Logistics systems, 103

Machine
    external, 55
    internal, 55
    language, 54, 55, 56
    man-machine, 86, 100
Machine operator, 95
Maintenance, 63
Management, 8, 49
    business, 134
    inventory, 135
    middle, 42
Management science, 71

Management game, 75, 80
Management Science, 124
Managerial control, 76
Manual
  game manager's, 75
  player's, 75
  programming, 75
  war gaming, 91
Market, 135
  information processing, 143
  mass, 60
  two-sided, 160
Mass data processing, 52
Mass experimentation, 75
Mass markets, 60
Mass society, 106
Master-pilot screen, 93
Mathematics, 5, 126, 168, 191
  sophistication in, 122
Matrix (*see also* games)
  payoff, 154
  three-person, 13
  3 × 3, 161
  2 × 2, 13
Maximum, 154
  point, 160, 163
Maxmin solution, 155
Maxmin strategy, 18, 161
Maxmin, the difference, 20
MBA curriculum, 141
Measure, ethicality, 159
Memory, high-speed, 57
Mental health, 60
Mental Health Research Institute, 126
Microeconomic theory, 139
Middle management, 42
Military, 6, 27, 125, 144
  large games, 23
Military analyst, 179
Military decision making, 173
Military exercises, 6, 7, 27
Military gaming (*see* games)
Military journals, 125
Military operations research, 144
Military training, 135
Minsky, M., 39, 133
Missile launching, 180
M.I.T., 167
Mode, binary, 55

Model, 7, 30, 103, 107, 124, 178
  air-battle, 180
  analog computer, 8
  building, 71, 100, 104, 106, 108, 122, 141
  construction, 51
  mathematical, 22, 140, 167
  military, 107
  rational, of man, 22
  traffic flow, 8
  war-game, 178
Modern period, 179
Money, 42, 160
  ransom, 16
  stakes, 46
Monitor, television, 86
Monopoly, 29, 44, 80
Monopoly
  behavior, 162
  bilateral, 153, 154, 157
Monte Carlo, 42
  method, 51
Motivation, 11, 30, 38
Mountaineering, 46
Murphy's law, 110
Musket period, 179

Nash solution, 160
National Gaming Council, 126
Naval Electronic Warfare Simulation
  (NEWS), 62, 83, 85, 91, 93, 94, 179
Naval Logistics Research Quarterly, 125
Naval Personnel, 126
Naval War College, 83, 91, 126, 179
Negotiation, 173
Networks, time-sharing, 57
Newtonian mechanics, 32
Nonconstant sum games, 18, 29, 31, 39
Noncooperative equilibrium, 19, 162
Noncooperative equilibrium solution, 17, 18, 24
Node, 12
Noise, 111
Normal form, 11
Normative solution, 18, 24
n-person, 24, 32
Nucleolus, 158, 163, 164
Numerical scale, 124

Objective function, 52
Observation gallery, 95
Office supplies, 7
Oligopoly, 154
Operations
    machine, 95
    military, 144
Operations, gaming (see Gaming)
Operations research, 3, 173
Opportunity costs, 68, 69
Output, 26, 110
    storage of, 75

Parades, 41
Parameter, 108
Pareto optimal, 157, 160
Payoff, 9, 11, 12, 17, 22, 44, 50
Payoff function, 181
Payoff matrices, 17, 154
PDP-5, 88
PDP-8, 88
Pearl Harbor, 37
Pecking order, 121
Pennsylvania State University, 157
Pentagon, 167
Period
    modern, 179
    musket, 179
    shock, 179
Personality, 25, 39
Personnel relations, 135
Physical science, 37
Physics, 168
Picture inputs, 54
Pilot, experimentation, 40
Pilot testing, 42
Planning, 33, 35, 36, 49, 50, 53
    economic, 140
    game, 35
    policy, 174
Play, 3, 5, 24, 29, 34, 40, 63, 64, 81
    84
    artificial, 38, 39
    chess, 7
    motivation, 38
    poker, 41
    robot, 160
    role, 25, 41, 45, 102
Point
    saturation, 107

saddle, 161
salient, 165
Poker, 7, 12, 22, 41, 44, 182
Poker player, 41
Policeman, 31
POLIS Laboratory at Santa Barbara,
    83, 88, 89, 90, 91
Political-military exercises, 102, 109,
    112, 167, 174
Political-military games, 173
Political science, 71, 89, 125, 144,
    167, 168, 179
    gaming in, 24, 125, 130, 166
Posture, 102
Power, 17, 32, 158
Prediction, 18
    contingency, 35
Prescription, 18
Prestige, 17
Price, 162
    competitive, 163
    variability in, 153
Priorities, 102
Prisoner's dilemma, 20, 25, 84, 154
Private corporations, 27
Problem solving, 39, 56
Processes, 56
    batch, 84
    random, 51
Production, 32, 76, 162
    inventory, 8
Professional standards, 26
Programming, 57, 95
    integer, 32
    languages of, 54, 78
    listing, 75
Programming manual, 75
Programming skill, 72
Pseudomonopoly, 162
Psychiatrist, 22, 43, 45, 114, 184
Psychoanalysis, 60
Psychology, 22, 94, 157
    aspects of, 182
    experimentation in, 37, 112
    social, 3, 13, 22, 45, 114, 115,
        144
Psychotherapy, 60
Purdue Laboratory, 83
Purpose, unstated, 101
Puzzles, 76

Puzzles (*cont'd*)
  crossword, 44, 101
  jigsaw, 44

Quarterly Journal of Economics, 125
Quid pro quo, 16

RAC, 86
Rand Corporation, 35, 86, 125, 167,
  172, 174
  Logistics Simulation Laboratory,
  86
Random process, 51
Random variables, 110
Ransom money, 16
Rationality, 11
  behavior, 22, 32
  group, 24
  individual, 24
  social, 24, 39
Rational model of man, 22
Real time, 93
Realism, 41
Receipts, box office, 101
Redevelopment, urban, 71
Reduction, data, 79
Referee, 23, 120, 121
Rehearsal, 33
Relevance, 104
Replacement, 63
Research, 61
  institute for, 69
  scientific, 120
Research Analysis Corporation
  (RAC), 125
Resolution, conflict, 125, 156
Resource constraints, 102
Response, surface, 79
Revenue, 70
Revolutionary strategy, 182
Rigid rule, 10
Risk
  calculated, 41
  extreme, 43
Risk of life, 42
Rivalist, 154
Robot, 40, 160
Role playing, 25, 41, 45, 102
Roulette, 41
Rome, 183

Rules, 11, 22
Rules of the game, 25

Saddle point, 161
Sage Publications, 124
Salient point, 165
Sand table, 6, 8
Santa Barbara
  Economics Laboratory at, 83
  POLIS Laboratory, 83
Saturation point, 107
Scale, numerical, 124
Scenario, 31, 81
  design, 181
  writing, 181
Scheduling, 178
  production inventory, 8
Scholarship, 129
Scottish Parliament, 183
Screen, master-pilot, 93
Searching, 39, 58
Senior ranking officer, 23
Sensitivity, 105
  analysis, 79, 107, 108
  search, 51
Set
  bargaining, 158, 163
  searching, 58
  stable, 158, 163
Shakedown cruise, 33
Shakespeare, 183
Shock period, 179
Shop, job, 109
Side payment, 16, 17, 20
Sierra series, 35
SIMCOM, 39, 85
Simplicity, 7
SIMSCRIPT, 56
Simulation, 3, 86, 107, 124, 126,
  167, 172, 178
  background, 3
  computer, 138
  experimental, 53
  exploratory, 8
  languages of, 138
  strategic, 8
  tactical, 8, 109
  technical, 109
Simulation planning, 53

Simulation and games, 124
Simulation councils, 125
Small-group behavior, 86
Smell, 54
Smithsonian Institute, 94
SMOG, 29
Snakeoil salesman, 36
Soccer, 43
Social contact formation, 45
Social psychology, 3, 13, 22, 45,
  114, 115, 144
Social status, 158
Society, mass, 106
Sociologist, 22, 89
Sociometrics, 49
Solitaire, 44
Solitary games, 44
So Long Sucker, 45
Solution, 17
  behavior, 17
  concepts, 11
  cooperative, 17, 18, 163
  dynamic, 17
  egalitarian, 164
  maxmin, 155
  Nash, 160
  noncooperative, 17, 18, 24
  noncooperative equilibrium, 19
  normative, 18, 24
  problem, 39
Soundproofing, 87
Sources, secondary, 120
Soviet literature, 126
Space
  flexibility of, 88
  floor, 87
Spare parts, 6
Specialist, logistics of, 94
Specification, 26, 27, 38, 103, 112
Spectacle
  mass, 40
  public, 40
Spectator sports, 43, 101
Sponsor, 4, 29, 34, 100, 101
Sports, 6, 27
  organized, 76
  spectator, 43, 101
Sportscar racing, 46
Stability, 107, 158, 163
Stanford Research Institute, 86, 125

167
Statistics, 49
  nonparametric, 114
  problems in, 114
Statistical tests, 79, 113
Status, social, 158
Steeplejack, 42
Storage, output, 75
Strategic form, 11
  gaming, 166
Strategy, 9, 13, 22, 33, 35
  maxmin, 18
  revolutionary, 182
Structure
  force, 86
  game, 184
Student, 33
  high school, 102, 103
Studies, Analysis and Gaming
  Agency, 30
Subroutines, 75
Summit, 29, 80
Supplies, 61
  office, 74
Surface response, 79
Symmetry, 105, 106, 116, 163, 164
  external, 25, 106
  internal, 106
Systems
  accounting, 70
  associative, 120
  audio, 87
  behavior, 113
  clearing-house, 86
  complex, 30
  computer, 56, 78
  defense, 73
  design, 94
  logistics, 103
  video, 87
  weapons, 34, 36
Systems Development Corporation
  (SDC), 62, 78, 83, 86, 88

Tactical air war, 180
Tactical tank battles, 22
Tactics, 125
"Taken for a sucker," 43
Tax income, 70

207

Tax laws, 73
Taxonomy, 100
Teaching games, 3, 27, 28, 29, 32, 84
Team control, 81
Technology, 57
Television, 59
Television monitors, 86
Temper, 174
Tennis, 44
Termination effects, 112
Testing, 33, 76, 116
  pilot, 42
  statistical, 79, 113
  Turing, 116
T-group, 45, 141
Theater, 26, 40
  Czech experimental, 31
  rehearsals for, 33
The General, 125
Theory, 112
Therapy, 3, 28
  group, 45
Threats, 16
Three-by-three matrix, 161
Three-person coalitions, 160
Three-person games, 161
Time
  real, 93
  turnaround, 84
Time-series data, 50
Time-sharing, 76, 78, 80, 96
Time-sharing networks, 57
Topology, 123
Touch, 54
Traffic, 85
Training, 3, 27, 28, 32
  military, 135
Training games, 32, 33, 85
Transaction flows, 56
Trial runs, 74
Triopoly, 154
Turnaround, 84
Tutoring, individual, 59
Two-by-two gaming, 37, 38, 45, 97,
  155, 156, 162, 167, 173
Two-person zero-sum games, 18, 29,
  39, 146
Two-person cooperative game, 165
Two-sided market, 160

U.C.L.A., 78, 88
Uncertainty, 114
Undergraduates, 37
Universities, 85, 166
Unknown soldier, 121
Urban development, 49
Urban planning, 24
Urban redevelopment, 71
User-written language, 55
U.S. General Accounting Office, 180

Value, 21, 163, 164
Variable, random, 110
Variance
  analysis of, 114
  price, 153
Verbal interaction, 22
Verification, 115
Video systems, 87
Voice, 54
Voting behavior, 166

Waiting line, 8
Walrasian Hypothesis, 164
War, 7, 182
  air, 180
  atomic, 7
  diplomatic-military games, 34
War college, 85
War exercises, 40
War game director, 178
War games, 23, 27, 37, 70, 86, 91,
  98, 126, 131, 178, 179
War Gaming Agencies, 30
War gaming amateurs, 30
War Gaming Department, 94
Weapons
  evaluation, 26, 58, 86
  system, 34, 36
Wei-Ch'i, 171
World, 37
World Politics, 125
World Politics Simulation (WPS), 85
Writing, 54

Yale, 78

Zero-sum games, 18, 29, 39, 146, 165
Zeuthen-Nash theory, 165